# Durability of geosynthetics

## CRC Press
Taylor & Francis Group
Boca Raton  London  New York

CRC Press is an imprint of the
Taylor & Francis Group, an **informa** business

A BALKEMA BOOK

# Durability of Geosynthetics

Published by:
CRC Press/Balkema
Schipholweg 107C, 2316 XC Leiden, The Netherlands

First issued in paperback 2023

ISBN 13: 978-0-367-73731-3 (pbk)
ISBN 13: 978-90-5367-599-1 (hbk)

It is allowed, in accordance with article 15a Netherlands Copyright Act 1912, to quote data from this publication in order to be used in articles, essays and books, provided that the source of the quotation, and, insofar as this has been published, the name of the author, are clearly mentioned.

**Liability**
SBRCURnet and all contributors to this publication have taken every possible care by the preparation of this publication. However, it cannot be guaranteed that this publication is complete and/or free of faults. The use of this publication and data from this publication is entirely for the user's own risk and SBRCURnet hereby excludes any and all liability for any and all damage which may result from the use of this publication or data from this publication, except insofar as this damage is a result of intentional fault or gross negligence of SBRCURnet and/or the contributors.

|  | Hartmut F Schröder PhD, consultant BPHS |
| --- | --- |
|  | Wim Voskamp MSc, Voskamp Business Consultancy |
| Editor: | Aad van den Thoorn, SBRCURnet |
| Book & cover design: | Annet Zorge & Linda de Haan, SBRCURnet |

# Contents

# PREFACE

CUR Building & Infrastructure in the Netherlands has identified the need for a publication on durability of geosynthetics. It is focused on the designer, the quality controller and supervising engineer on site and the owner of a project, and should also be of use to other interested parties such as manufacturers, testing institutes and students.

This publication addresses the various mechanisms by which geosynthetics degrade, lists tests which could be specified to measure the long term effects, and the methods by which the tests can be evaluated in order to determine the expected lifetime of a geosynthetic. Individual chapters describe the types of mechanism, based on the requirements for 100 years service life.

The present Harmonised EN standards do not help the engineer who has to design or approve a geosynthetic material for a service life of more than 25 years. There is a need for instructions on how to solve this problem and to identify the experts who can make the assessments.

The publication is intended for use by civil engineers who have a limited knowledge of chemical engineering or polymer technology, but who have to check the service life and are aware of the degradation of geosynthetics and also have knowledge of design with reduction factors, for example with reinforcement applications. We have tried to prevent as much as possible the use of chemical and polymer formulations, formulae and equations in the text. In some cases it was not possible to prevent it, in other cases an attachment was made with a basic chemical description.

At the same time the publication gives a state-of-the-art report of the various methods which are available to test and assess the life-limiting mechanisms. It describes the evaluation methods and describes how to make an assessment. This assessment can only be made by experts. The names of accredited testing institutes and certification bodies are given.

It covers the use of geotextiles, geogrids, drainage mats etc. in most civil engineering applications (see chapters 1.3 and 2.2 for more information). However it does not cover the use of geomembranes and geosynthetic clay liners, the application of which is in a different market segment. Furthermore geosynthetics of biodegradable material are not discussed, because the lifetime of such materials is not important - they are made to degrade quickly!

This publication was prepared under the responsibility of CUR committee C187 "Durability of geosynthetics".

Section 1 has been written by W. Voskamp MSc.
Section 2 and 3 have been written by Dr. J. H. Greenwood
Dr. H.F. Schroeder is the co-author of chapters 3.1 and 3.2

For chapter 3.3. the co-authors are Dr. H.F.Schroeder and Dipl.-Phys. P. Trubiroha

Sections 2 and 3 are based on seminars on "Life prediction in geosynthetics" held in Leatherhead, Würzburg and Dubai over the period 2000-2008.

At the time of issue of the publication, this CUR committee consists of the following members:

| | |
|---|---|
| Dr. J. H. Greenwood, author | Consultant to ERA Technology Ltd. |
| Dr. H.F. Schroeder, author | Bundesanstalt für Materialforschung und – prüfung (BAM), Consultant BPHS |
| W. Voskamp MSc., author and editor | Voskamp Business Consultancy |
| E. Berendsen MSc. | Rijkswaterstaat Dienst Infrastructuur |
| Dipl.-Ing. A. Elsing | Huesker |
| D. den Hamer MSc. | Deltares/on behalf of NGO |
| Dipl.-Ing. H. Homölle | PROPEX Fabrics |
| H.A. Jas MSc. | Tensar |
| Prof. Dr.-Ing. J. Müller-Rocholz | KIWA TBU GmbH |
| D.P. de Wilde MSc. | Rijkswaterstaat Dienst Infrastructuur |
| Ing. A. Jonker, coordinator | CUR Bouw & Infra |

The final editing of this publication was done by W. Voskamp.

Dr. rer. nat. M. Böhning and Dr. rer.nat. D. Robertson, Bundesanstalt für Materialforschung und -prüfung, BAM, Berlin have supplied valuable information, diagrams and photographs for chapter 3.1.

This publication has been peer reviewed by a panel of experts and is finally approved by the CUR 187 committee.
Some of them have reviewed one or more chapters, others have reviewed the complete publication.
We thank them for their detailed comments and valuable suggestions.

Members of the peer reviewing group:
Prof. Richard J. Bathurst, P.Eng., Ph.D., FEIC, FCAE, Kingston, Canada

Prof. Dr.-Ing. J. Müller-Rocholz, Germany

Dr.-Ing. J. Retzlaff, Germany

H. Jas, MSc

E. Berendsen, MSc

SBRCURnet would like to thank the organizations and companies listed below, whose support made this publication possible:

Rijkswaterstaat Dienst Infrastructuur

NGO

Port of Rotterdam

Deltares

Huesker Synthetic GmbH

KIWA TBU

Propex Fabrics GmbH

Tensar International B.V.

June, 2012 The board of CUR Building & Infrastructure

# SECOND EDITION

Chapter 4 on geomembranes and geosynthetic clay liners was added in this second edition. Comments which were received on the first edition have been included and some typing errors were corrected.

The peer reviewing group has been enlarged with:
Dr.Ing. D. Cazzuffi

Dr. G. R. Koerner

A. Needham, MSc, CEng, FICE, FGS

Dr. I.D. Peggs

SBRCURnet would like to thank everyone involved with the preparation of this second edition, for the work they have done.

March 2015, The Board of SBRCURnet

# GUIDANCE FOR THE USE OF THE REPORT

The first section of the report is to be used by engineers, supervisors and owners who are interested in the general requirements of durability aspects or who need to specify and approve material based on a required service life.

It contains a general introduction of geosynthetics, a description of the various steps in design focussed on durability and the determination of the service life of geosynthetics. Practical information is given and steps identified for the engineer and supervisor to follow to determine the service life. Detailed information is given in attachments, including the design procedure and the quality control procedure.

Chapter 1 contains:
- A brief introduction of geosynthetics.
- It is followed a chapter which describes why it is necessary to predict the lifetime of geosynthetics.
- Design aspects and material properties. Various subchapters describe:
  - Failure mechanisms
  - Detailed design of geosynthetics
  - Various design aspects and material aspects
- Durability aspects for design
  - Reduction factors for reinforcement; and drainage and filtration
  - Maximum lifetime < required service life
- Durability assessment
  - Required service life up to 5 years
  - Required service life up to 25 years
  - Assessment of long term durability (more than 25 years)
    - Approval procedure during tendering (focussed on durability aspects)
    - Assessment certificate
    - Verification of supplied material
    - Expert assessment institutes
- Quality control aspects
- Guide to establish the long term properties of geosynthetics based on durability

| | |
|---|---|
| Attachment A: | Design procedure and design of the geosynthetic. |
| Attachment B: | List of comparable EN, ISO and ASTM test methods |
| Attachment C: | Steps to be taken to assess the durability in accordance with prEN 13249 – prEN 13257 version 2011. |
| Attachment D: | Accredited test and certifying bodies |
| Attachment E: | Quality control according to EN standards |
| Attachment F: | Examples of CE documents |
| Attachment G: | List of relevant standards |

Thesecond sectionof the report contains more detailed information.
It describes the applications and types of geosynthetics in detail. It is followed by an in-depth description of all durability aspects, including a description of the mechanism, the test methods, extrapolation methods etc. This section is for test laboratories and super-vising bodies and manufacturers, who want to know all the aspects. It can be used as a description of the present knowledge.

Chapter 2 gives:
- Detailed introduction in geosynthetics, polymers and additives
- Various applications and related durability requirements
- Environmental aspects
- General durability aspects and introduction to
  - Required and available properties
  - Extrapolation of measured / tested data
  - Reduction factors
  - Accelerated testing
    - By increasing frequency
    - By increasing severity
    - By increasing temperature (Arrhenius' formula)
  - Uncertainty of extrapolation
  - Changes in the state of material
  - Index tests
  - Limits of prediction

Chapter 3 addresses the various degradation mechanisms:
- Oxidation
- Hydrolysis
- Weathering
- Tensile creep
- Compression creep
- Mechanical damage

- Method of life prediction
  - Reinforcement
    - Creep
    - Calculation of reduction factor for installation damage
    - Calculation of reduction factor for weathering
    - Calculation of reduction factor for chemical degradation
    - Factor of safety regarding reduction factors
    - Residual strength
  - Drainage, Filtration
    - Reduction factor for instantaneous compression
    - Reduction factor for instantaneous intrusion by the soil
    - Reduction factor for time-dependent compression of the core (compressive creep)
    - Reduction factor for time-dependent intrusion of the soil
    - Reduction factor for chemical degradation
    - Reduction factor for particulate clogging
    - Reduction factor for chemical clogging
    - Reduction factor for biological clogging
  - Data and extrapolation of exhumed material
  - Remanent life assessment

18

Contents

# CHAPTER 1
# INTRODUCTION TO DURABILITY ASSESSMENT

## 1.1    Introduction

Geosynthetics include a variety of synthetic polymer materials that are specially fabricated to be used in geotechnical, geo-environmental, hydraulic and transportation engineering applications.
With few exceptions geosynthetics consist of polymers, whether in the form of textile fibres, extruded sheet, coatings or more clever characterizations.

The International Geosynthetics Society lists as *main application areas*:
• Unpaved Roads
• Road Engineering
• Railroads
• Walls
• Slopes over Stable Foundations
• Embankments on Soft Soils
• Landfills
• Wastewater Treatment
• Hydraulic Projects
• Drainage and Filtration
• Erosion Control
• Agricultural Applications.

These application areas are described in much detail in chapter 2.2

It is convenient to identify the *primary function* (based on the EN standards) of a geosynthetic as being one of:
• separation (soil layer separation)
• filtration (soil retention and cross-plane water permeability)
• drainage (in-plane water discharge)
• reinforcement (soil strengthening)
• containment (ballast and structural elements)
• hydraulic barrier (lining and leakage prevention)

For practical reasons we add to this list in this report, in line with earlier CUR and PIANC reports:
• erosion control (surface soil wash-out or scour at structure foundations).
• protection (prevention of damage to the geosynthetic)

In some cases the geosynthetic may serve multiple functions.

The erosion protection function, although it in fact is a combination of 2 functions, is added here as an important application of geosynthetics in waterfront and marine applications.

The term containment is used to describe the encapsulation of sand into geotextile bags, containers or tubes. Sometimes the term confinement is used for this application, but the term containment is mostly used in the literature and therefore also used in this report.

The *types of geosynthetics* can be divided in:
* Geotextile
    * Woven material
    * Knitted material
    * Nonwoven materials
        * Thermally bonded
        * Mechanically bonded (needle punched)
* Geomembrane
* Geosynthetic Clay Liner
* Geogrid
* Geonet
* Geocell
* Geomat
* Geocomposite
* Geofoam

Details of these geosynthetics are given in chapter 2.1.

They can be made of various types of polymers, each of which have different characteristics and will therefore influence the performance of the products.

Geosynthetics are manufactured from the following *main types of polymeric materials*:
* Polyester (PET);
* Polypropylene (PP);
* Polyethylene (PE);
* Polyamide (PA);
* Polyvinyl alcohol (PVA).

No geosynthetic product is made of 100% polymer resin. The primary resin is mixed, or formulated with antioxidants, screening agents, fillers and/or other materials for a variety of purposes. The total amount of each additive in a given formulation varies widely. The additives function as UV light absorber, antioxidants, thermal stabilizers, etc.

The types of polymers and their characteristics are described in chapter 2.2

*Material aspects*
* PP and PE are mostly used for filtration materials.

- PET is specifically suitable for reinforcement applications, because the creep is low.
- Nonwoven and woven fabrics can be used for filter and separation applications because the tensile loading in these cases is limited.
- Typically a geogrid made of extruded PE, strips containing PET fibres or a heavy PET woven fabric is used in reinforcement applications when high tensile loadings must be taken up at low strain.
- Nonwovens are, due to thickness and elongation of this material, often used for filtration, containment and protection of barrier applications.
- High density Polyethylene (HDPE) and Low Density Polypropylene (LDPE) are used as barrier materials.

According to Koerner (2005) the *main advantages of geosynthetics* are:
- They are quality-control manufactured in a factory environment
- They can be installed rapidly in many cases
- They generally replace natural resources
- They generally replace different designs using soil or other construction materials
- They are generally cost competitive against the soils or other construction materials that they replace or augment
- Their technical database (both design and testing) is reasonably established

Geosynthetics can be used in different *environmental conditions*:
- In the ground
- Above the ground
- Under water
- Above water
- At the level of the water table

These environmental conditions influence the projected service life of the geosynthetic. Factors of influenceare:

- Temperature of the soil
- Rainfall
- Type of soil and particle size
- Method of installation
- Chemical composition of the soil
- Intensity of UV

These conditions and their effect on the projected service life of a geosynthetics are described in detail in chapters 2.2 and 2.3.

The various degradation mechanisms are described in chapter 3. They are important for the determination of the service life of a geosynthetic. A sufficient service life or lifetime of a geosynthetic is essential for the safety of the structure in which it is used. The design aspects are discussed later chapters 1.3 and 1.4.

## 1.2     Why predict lifetime?

Definitions:
Required / Design life or lifetime is the required time during which the geosynthetic must possess all required properties in the specified conditions. Sometimes the term required service life is also used for this design requirement.
Expected or projected service life is the lifetime of the material during which the geosynthetic is expected to possess all required properties in the specified conditions to guarantee the design factors of safety, based on a durability assessment.
Lifetime assessment is an assessment made by experts of the projected service life of a geosynthetic.

*General*

Most commonly a manufacturer, design engineer or installer has to demonstrate to the buyer or final owner of a structure that, as far as anyone can predict, the geosynthetic will last for the projected service life.

The projected service life can be the same as the required lifetime of the structure, for example in the case of a:
*   roadway
*   reinforced wall
*   dam
*   tunnel
*   landfill.

Not all structures depend on geosynthetics for their whole lifetime. In some cases the geosynthetic has a temporary function, such as:
*   the stabilization of a dam against slip failure until the soil has consolidated
*   an embankment over a weak soil which requires reinforcement only until the soil has compacted
*   in a temporary (access) road
*   vertical drainage (PVD) under an embankment to increase the consolidation of the subsoil.

To illustrate this some examples of contrasting design lifetimes are given in the box below.

| Separator as construction aid | 0.5-1 year |
|---|---|
| Separator (permanent) | 80-100 years |
| Filter in drainage applications, replaceable | 10-25 years |
| Filter in drainage applications, not replaceable | 80-100 years |
| Reinforcement in a dam against slip failure | 5 years |
| Reinforcement in retaining structures | 80-100 years |
| Prefabricated vertical drain | 1-3 years |
| Landfill liner, drainage and protection | 100+ years |
| Tunnel liner | 100+ years |

*Degradation of polymers*

Geosynthetics can be damaged even before they start their proper function, due to:

- production failures (extended time in extruder and/or temperature differences)
- mechanical damage during transport
- transport at too high temperatures for longer periods (containers to or in tropical countries)
- mechanical damage during loading, unloading and storage
- unprotected storage for too long time
- uncovered for too long time
- weathering
- installation damage

The principal causes of degradation for geosynthetics covered and buried in the ground are:

- Mechanical load can cause long term extension known as creep. Dynamic loads due to traffic can lead to rupture known as fatigue. Earthquakes cause intense mechanical stresses.
- Slow chemical attack occurs due to oxygen, water and dissolved components in soil. Polypropylene and polyethylene contain special chemicals called antioxidants which prevent oxidation, but in the long term these can be consumed, dissolve, evaporate or be rendered ineffective, after which oxidation sets in. In polyesters the water leads to a slow decomposition by means of hydrolysis, while in polyamides both oxidation and hydrolysis can occur.
- PVC geomembranes contain plasticizers, the loss of which over time leads to embrittlement and shrinkage, causing particular problems at connections.
- Drainage materials can compress, restricting the flow of liquid
- Geosynthetic clay liners can degrade due to shear forces and chemical changes in the clay as well as degradation of the geotextiles within them. The clay can dry out and then rehydrate. (not further discussed in this report).
- Some geosynthetics can be affected by biological effects such as rodents, plant roots and microorganisms.
- In certain applications and regions high energy radiation can cause changes to the polymer.

In this report (chapter 2) we shall describe those forms of degradation that can themselves be predicted, the rate at which they occur, and how an understanding of this can lead to methods for predicting lifetime. We cannot predict earthquakes or rodent attack, where the choice of a suitably resistant geosynthetic is a matter for design rather than life prediction. We can only include the degradation of the geosynthetic itself and not problems such as clogging which are due to the manner in which it is used. The effects of high energy radiation on polymers are a specialist subject which will not be treated here.

*Methods for life prediction*
Most commercial geosynthetics have in fact proved themselves durable in the fifty years since they were first installed, and over that time the quality of the products available has been improved. This experience, coupled with an understanding of the physics and chemistry of degradation, has led to the introduction of screening or index tests.

Methods of life prediction can be based on:
- Direct measurements on exhumed material, extrapolated in time (so far information has only been available for short durations).
- Results of simple (index) tests, which can be performed within a reasonable time. They can provide an assurance that a conventional geosynthetic will last for at least 25 years in a 'normal' soil environment. No conclusion can be drawn for a service life of more than 25 years.
- In case a lifetime of more than 25 years is required, generally the only way to obtain sufficient information is to execute accelerated testing. The assessment of the projected service life based on these results can only be made by experts.

This more formal and comprehensive life assessment is required in case of a geosynthetic to be used outside these limits:
- for new or different polymers
- at higher temperatures
- in more aggressive environments
- for all for service lives more than 25 years.

This report is intended to set out how to perform such an assessment.

To make such an assessment requires a precise definition of the geosynthetic and the proposed environment. It also requires knowledge of the types of degradation that can occur.

The durability of a geosynthetic, i.e. its resistance to chemical degradation, will depend on:
- The type of polymer
- The macroscopic structure of the geosynthetic, such as the thickness of geo-membranes and the diameter of fibres

- The purity of the raw materials used and the presence of impurities
- The polymerization process and the catalysts used, if any
- Copolymerization, which generally reduces the durability
- The processes used to create the geosynthetic from the basic polymer
- The storage of the raw materials before and after processing
- The additives, as described in chapter 2.1.3

While the type of polymer and the structure of the geosynthetic will form part of the selection process, the user will have no control over the other aspects. Durability assessment is therefore based on tests on the finished product, together with appropriate quality control measures.

One point however needs stating, and will be repeated again and again in this report. Any prediction of the service life of geosynthetics, is totally dependent on correct design, selection and quality of installation, including joints. Without that assurance you need read no further.

## 1.3    Design aspects and material properties

A detailed description of the design procedure based on a system engineering approach, as used by the Dutch Department of transport (RWS), is given in attachment A.

The main phases of the design procedure (or technical process), referring to figure A.1.1 are:
1. Requirement analysis (clarification of the task by collecting the information about the requirements and the project constraints; requirements can be customer, functional (what), performance (how many, how far, when and how long and how often), design (e.g. how to execute), requirements)
2. Translation of the requirements into a coherent description of system functions
3. Design synthesis (the search for suitable solution principles and their combinations into concept variants/design alternatives):
   - Conceptual
   - Detailed

The detailed design includes a safety analysis. Safety requirements (for example the use of safety factors) are used to determine the required minimum long term values of properties of the geosynthetic. For example the long term design strength of the selected product must have properties which fulfil these requirements. This must be verified during execution.

### 1.3.1    Detailed design

Design rules and calculation methods are available for many functions in various applications:
- Filtration
- Separation
- Permeability
  - Geotextiles
  - Geomembranes

27

- Water flow (capacity) in the plane and drainage
- Soil reinforcement
- Seams and overlaps
- Hydraulic Barrier
- Containment

These design rules and calculation methods can be found in the CUR reports 151, 174, 206 and in the PIANC report 113 or in other reference textbooks.

## 1.3.2    Design aspects and characteristic material aspects

Table 1.1 gives an overview of the design aspects in relation to the functions of the geosynthetic. The design aspects are divided into material characteristics or properties. A property or material characteristic is indicated when it is important for a function.

Table 1.1 indicates if a characteristic must be specified (x) or must only be specified in special cases resulting from the design (0). In case the characteristic is not relevant for the specific function it is indicated as (-) and does not have to be specified. Full details about the background of the table can be found in the latest version of the EN application standards.

The relevant EN standards and some relevant ASTM standards are mentioned. There are also other standards which can be used to determine the characteristic design aspects and material properties. A list of comparable EN, ISO, ASTM test methods and standards to determine the characteristic design aspects is given in Attachment B

Table 1.1 Materials properties which are relevant to be specified for the indicated functions, based on application standards EN 13249 – 13257, EN 13265 and EN 15381.

| Design aspect | Characteristic material aspect | Test-method | Function | | | | | | |
|---|---|---|---|---|---|---|---|---|---|
| | | | Filtration | Seperation | Drainage | Reinforcement | Barrier | Erosion protection | Confinement |
| **Physical design** | **Physical Properties** | | | | | | | | |
| | 1 Thickness | EN 1849-2 | - | - | - | - | x | x | 0 |
| | 2 Mass per unit area | EN 1849-2 | - | 0 | - | - | x | x | 0 |
| **Hydraulic design** | **Hydraulic Properties** | | | | | | | | |
| Retention | Characteristic opening size O90 | EN ISO 12956 | x | x | x | - | - | x | x |

| | | | | | | | | | |
|---|---|---|---|---|---|---|---|---|---|
| Water permeability normal to the plane | Water permeability normal to the plane VI50 | EN ISO 11058 | x | x | x | - | - | x | o |
| Water perme- ability) | Water permeability (liquid tightness) | pr EN 14150 | - | - | - | - | x | - | - |
| Waterflow capacity in the plane | Waterflow capacity in the plan (flow rate q at specified loads and gradients) | EN ISO 12958 | - | - | x | - | - | - | - |
| **Mechanical design** | **Mechanical Properties** | | | | | | | | |
| Mechanical strength | Tensile strength | EN ISO 10319 | x | x | x | x | - | x | x |
| | | ISO R 527 | - | - | - | - | x 1) | - | - |
| | Burststrength | prEN 14151 | - | - | - | - | o 1) | - | - |
| | Tear strength index | ISO 34 | - | - | - | - | o 1) | - | - |
| Strength of seams and joints | Tensilte strength of seams and joints | EN ISO 10321 | o | o | o | o | x | - | x |
| Deformation | Elongation (at maximum load) | EN ISO 10319 | x | x | x | x | - | - | x |
| | | ISO R 527 | - | - | - | - | x 1) | - | - |
| | Static puncture (CBR test) | EN ISO 12236 | o | x | o | x | - | - | x |
| | | EN ISO 12236 | - | - | - | - | x | - | - |
| | Dynamic perforation resistance (cone drop) | EN ISO 13433 | x | x | o | x | - | - | x |
| Stability | Friction Direct shear | prEN ISO 12957-1 | o | o | o | x | o | o | o |
| | Friction Inclined Plane | prEN ISO 12957-2 | o | o | o | x | o | o | o |

| Long term strength | Tensile creep | EN ISO 13431 | - | - | - | x | - | - | x |
|---|---|---|---|---|---|---|---|---|---|
| Robustness / Installation aspects | Damage during installation | ENV ISO 10722-1 | x | x | x | x | - | o | x |
| | Protection efficiency | EN 14574 | - | - | - | - | - | - | o |
| **Thermal design aspects** | **Thermal Properties** | | | | | | | | |
| | Low temp behavior (flexure) | EN 495-5 | - | - | - | - | o | - | - |
| | Thermal expansion | ASTM D 696-91 | - | - | - | - | x | - | - |
| **Durability and Chemical Resistance** | **Durability and Chemical Resistance** | | | | | | | | |
| **Durability and Chemical resistance** | Durability according tot annex B | Annex B | x | x | x | x | x | x | x |
| | Resistance to weathering 3) | EN 12224 | x | x | x | x | x | x | x |
| | Resistance to oxidation and chemical ageing 3) | ENV ISO 12960 or ENV ISO 13438, ENV 12447 | x | x | x | x | x | x | x |
| | Resistance to microbiological degradation 3) | EN 12225 | o | o | o | o | x | o | o |
| | Environmental stress cracking 2) | ASTM D 539-99 | o | o | o | x | x | o | o |

1) Only applicable for geosynthetic barriers (PP, PE)

2) In case of a polyolefin polymer (PP, PE), PA only oxidation

3) Relevance of the design aspect is described in Annex B of the relevant application norms: EN 13253, 13254, 13255 and other relevant application norms

x = to be specified o = to be specified in special applications - = not relevant

## 1.3.3   Failure mechanisms

1.  Abrasion[*]
2.  Instability of geosynthetic elements[**]
3.  Washing away of fill material[**]
4.  Perforation of the geosynthetic elements by puncturing[*]
5.  Rupture of seams and joints[**]
6.  Occurrence of S- shaped profile under the filter layer in a slope[**]
7.  Degradation by UV radiation or other ageing processes[*]
8.  Damage during installation[*]
9.  Decrease of permeability by clogging and blocking[**]
10. Rupture of a reinforcing element[**]
11. Loss of anchorage of a reinforcing element[**]
12. Insufficient drainage of the structure[**]

[*] These aspects influence the service life of the geosynthetic and are discussed in this report.
[**] This has an effect on the service life of the geosynthetic but is a result of the manner in which it is used. Therefore it is not discussed in this report, more detailed infor-mation can be found e.g. in the CUR 174 report.

# 1.4   Durability aspects for design

## 1.4.1   General

One of the requirements on a geosynthetic is that certain minimum values for the main prop-erties are present during the entire service life of the structure.
The long term properties of geosynthetics are determined by the presence of certain additives which are mixed in the polymer during extrusion. These additives are applied in small quan-tities during production. This results in an expected service life for many geosynthetics of more than 25 years. This does not mean that these geosynthetics will survive also 100 years. More additives must be added during production to achieve that. This is costly and is not done for standard products. Products with an expected long lifetime are therefore mostly specially produced (to order). Reinforcement products on the contrary are normally produced for a service life of at least 100 years.

For a designer it is important to know that, within the normal range of required service life's (0 – 120 years), it is always possible to produce products which could fulfil these service life requirements. It is therefore during design not necessary to anticipate possible shorter service lives.

*Durability aspects*

The service life of a structure with a geosynthetic depends for a large part on the durability of the geosynthetic.

The properties of the geosynthetic can change in time as result of:
- Mechanical load (creep)
- UV radiation
- Chemical and biological attack, including oxidation and hydrolysis
- Mechanical damage, abrasion and dynamic loading

The durability effects on all major material characteristics must be taken into account.

In chapters 3.1 to 3.6 we describe the various mechanisms in detail and conclude with the method of service life assessment in chapters 3.7 and 3.8.

Table 1.2 gives the relevance of the various durability aspects as described in chapter 3 for the functions of a geosynthetic.

Table 1.3 indicates the method which is used to take care of the degradation mechanisms during the determination of the expected service life. This can be done by:
- application of a reduction factor to the (short term) index value
- the maximum service life will be given in case of a limited life time. A reduction factor on the index value will also mostly be applied.

The short term (index) value of the property is divided by the reduction factor to obtain the expected long term value.

This expected long term value of the property must be more than the required long term value of the property which was specified during the design.

---

In all cases it must be verified that the selected product fulfils the requirement:

**Expected long term value of the property > required long term design value of the property**

and in case the expected service life is limited:

**Expected service life > required design lifetime**

Safety factors have been included in the design, therefore it is not necessary to apply them in this calculation.

---

Reduction factor

During long term or accelerated tests, the reduction in the value of a certain property in time is determined. The reduction factor is calculated by dividing the (short term) index value of a property by the measured value at the end of the service life.

Chapter 1.5.6 gives default values for these reduction factors. These default values are based on an evaluation of data of many tests and of many products. These default values can be used in case no further data is available. It will be clear that use of default values leads to low long term property values, therefore it is advised to use as much as possible test data of products, or to use products for which assessments have been made.

While for most applications the purpose of life prediction is to ensure that the projected service life of the geosynthetic is greater than the required design lifetime, in some cases a reduction of the index test value of a geosynthetic property is allowed. This reduction is then calculated during the design and the reduced value is used as the long term property of the material. This is for example the case with strength in reinforcement applications or in some cases with the discharge capacity of drainage mats or permeability in filter applications. These methods are described in detail in sections 1.5.2 and 1.5.3 and chapter 3.7.

In other cases the use of a Reduction Factor is not appropriate to design service life, particularly in non-reinforcing applications. For example in the case of oxidation, when the anti-oxidants have been depleted, polymer degradation, and with it the reduction in mechanical strength, takes place within a period that is relatively short. In that case a limit must be set on the functional service life of the geosynthetics for example determined by a minimum retained strength (or any other governing parameter). No RF will be allowed, only a maximum service life will be declared. Refer to chapters 1.4.4, 2.3, 3.1, 3.7 and 3.8.

For some properties such as resistance of nonwovens to damage and the protection of geomembranes current practice is that durability is assured by a single index test or by selection of a suitable class of material.

Table 1.2 Relevance of the various durability mechanisms for specific geosynthetic functions.

| | Mecha-nism: | Oxidation | Hydrolysis | Weath-ering | Tensile creep | Compres-sion Creep | Instal-lation damage |
|---|---|---|---|---|---|---|---|
| | Chapter: | 3.1 | 3.2 | 3.3 | 3.4 | 3.5 | 3.6 |
| | | PP, PE 1) | PET 1) | | | | |
| Function | | | | | | | |
| Filtration | | o | o | + | - | - | + |
| Separation | | o | o | + | - | - | + |
| Drainage | | o | + | + | - | + | + |
| Reinforce-ment | | + | + | + | + | - | + |
| Barrier | | + | - | + | - | - | - |
| Erosion protection | | o | o | + | - | - | + |
| Confine-ment | | o | + | + | - | - | + |

1) only relevant in case the geosynthetic is made of these polymers.

+ means relevant for the function
o means relevant for the function in some applications
- means not relevant for the function

Table 1.3 Methods to include the durability aspects on the main properties for various mechanisms.

| Mechanism: | Oxidation | Hydrolysis | Weath-ering | Tensile creep | Compres-sion creep | Installation damage |
|---|---|---|---|---|---|---|
| Chapter: | 1.5.4; 3.1 | 1.5.4; 3.2 | 1.5.1; 3.3 | 1.5.6; 3.4 | 1.5.6; 3.5 | 1.5.6; 3.6 |
| | PP, PE 1) | PET 1) | | | | |
| Main pro-perty | | | | | | |
| Opening size | Approve and apply $RF_{CH}$/ reject | Approve and apply $RF_{CH}$/ reject | Limitation of uncovered use | Increased value if ap-plicable 2) | Apply $RF_{CC}$ | Increased value if ap-plicable 2) |
| Permeability | Approve and apply $RF_{CH}$/ reject | Approve and apply $RF_{CH}$/ reject | Limitation of uncovered use | Increased value if ap-plicable 2) | Apply $RF_{CC}$ | Increased value if ap-plicable 2) |
| Flow capa-city in the plane | Approve and apply $RF_{CH}$/ reject | Approve and apply $RF_{CH}$/ reject | Limitation of uncovered use | Check if $RF_{CR}$ is needed | Apply $RF_{CC}$ | Approve apply $RF_{ID}$/ reject |

| Mechanical Strength | Approve and apply $RF_{CH}$ / reject | Apply $RF_{CH}$ | Limitation of uncovered use or apply $RF_W$ | Apply $RF_{CR}$ | N/A | $RF_{ID}$ |
| Elongation | Approve and apply $RF_{CH}$ / reject | Apply $RF_{CH}$ | Limitation of uncovered use or apply $RF_W$ | Apply $RF_{CR}$ | N/A | $RF_{ID}$ |
| Puncture resistance | Approve and apply $RF_{CH}$ / reject | Apply $RF_{CH}$ | Limitation of uncovered use or apply $RF_W$ | Apply $RF_{CR}$ | N/A | $RF_{ID}$ |

1) only relevant in case the geosynthetic is made of these polymers

2) The opening size or permeability could increase as result of creep or installation damage; apply a RF>1 if relevant

Approve and apply $RF_{CH}$ /reject: Reject in case the maximum lifetime of the geosynthetic is less than the required design life of the geosynthetic; otherwise apply R

## *1.4.2  Reduction factors for reinforcement applications*

In the specific case of reinforcements a loss in strength is allowed for during the life of the reinforced soil structure. Strength is a critical property of reinforcement; without it the soil structure would collapse. This strength diminishes with time due to a variety of influences. To allow for this the initial (ultimate) strength of the geosynthetic - the manufacturer's proud and quality assured number – is reduced by a number of reduction factors reflecting the stress the geosynthetic is under and the environment surrounding it. Most of these factors will themselves change with time. ISO/TR 20432 defines them as follows:

$RF_{CR}$ : reduction factor for creep-rupture

$RF_{ID}$ : reduction factor for installation damage

$RF_W$ : reduction factor for weathering

$RF_{CH}$ : reduction factor for the environment, including chemical and biological attack

For applications such as railways, it may be appropriate to add a further reduction factor for dynamic loading (see sections 3.6.7 and 3.7.2.8). The German system adds a reduction factor for joints and connections (see section 3.7.2.8).

The reduction factors are greater than 1.0 or, if no degradation is expected, equal to 1.0. In addition there is a further factor of safety $f_s$ to take into account uncertainties. All these factors are multiplied together to yield a single factor by which the initial strength is divided to give the design strength.

The general reduction factor RF is calculated as:

$$RF = RF_{CR} \times RF_{ID} \times RF_W \times RF_{CH} \times f_s.$$

According to ISO/TR 20432 the characteristic strength of the geosynthetic is reduced by the following four reduction factors and one safety factor, with their Dutch, German, US and British equivalents:

Table 1.4 Reduction factors according to ISO/TR 20432 (ref. chapter 3.7.2.1).

| ISO TR 20432 | Netherlands, Germany | United Kingdom BS8006 | Form of degradation |
|---|---|---|---|
| | | $f_{m111}$ | variability of initial strength |
| | | $f_{m112}$ | metallic reinforce-ment only |
| $RF_{CR}$ | $A_1$ | $f_{m121}$ | creep-rupture |
| $RF_{ID}$ | $A_2$ | $f_{m211}$ (short term) $f_{m212}$ (long term) | installation damage |
| | $A_3$ | | joints and connec-tions |
| $RF_W$ | $A_4$ | $f_{m22}$ | weathering |
| $RF_{CH}$ (USA: $RF_D$) | | | chemical degradation |
| | $A_5$ | | Special conditions, e.g. dynamic loads |
| $f_s$ | $\gamma_M$ | $f_{m122}$ (creep extrapolation only) | |

Strength is not the same as stiffness. In many reinforcement applications, a limit is set on the maximum elongation of the geosynthetic during its service life. In such a case, in addition to the calculation of strength using reduction factors there must be a calculation of long-term strain in parallel. The design load may be limited by the restriction on strain as much as by those of rupture.

*Calculation of $f_s$*

The factor of safety $f_s$ included in the calculation of design strength in ISO/TR 20432 accounts solely for the following:
- uncertainty of extrapolation of creep-rupture data ($R_1$)
- uncertainty of extrapolation of Arrhenius data ($R_2$).

These are explained in more detail in the following sections.

The factor $f_s$ depends on 2 factors: $R_1$ and $R_2$.

$$f_s = 1 + \sqrt{\{(R_1 - 1)^2 + (R_2 - 1)^2\}}.$$

$R_1$ depends on the duration of the longest creep test, $t_{max}$. If the creep data are extrapolated by less than $10 \times t_{max}$ then $R_1 = 1.0$. For extrapolation to design lives $t_d$ between $10 \times t_{max}$ and $100 \times t_{max} R_1$ increases with log ( $t_d / t_{max}$ ) from 1.0 to 1.2. Further details are given in section 3.7.3.3.

Methods for the calculation of $R_2$ are given in section 3.7.3.4.

When default values are used for $RF_{CH}$, $R_2 = 1,0$ according to ISO TR 20432 and $f_s = 1$.

## 1.4.3   Reduction factors for Drainage and Filtration applications

The same approach with reduction factors can be used in other applications. For drainage the main property is flow in the plane, and the factors proposed are described below and in section 3.7.7.

A method for predicting the reduction in flow due to compression creep is described in 3.5.

Basically the following Reduction Factors are used:

| $RF_{IMCO}$ | Reduction factor for instantaneous compression |
|---|---|
| $RF_{IMIN}$ | Reduction factor for instantaneous intrusion by the soil |
| $RF_{CR}$ | Reduction factor for time-dependent compression of the core (compressive creep) |
| $RF_{IN}$ | Reduction factor for time-dependent intrusion of the soil |
| $RF_{CH}$ | Reduction factor for chemical degradation |
| $RF_{PC}$ | Reduction factor for particulate clogging |
| $RF_{CC}$ | Reduction factor for chemical clogging |
| $RF_{BC}$ | Reduction factor for biological clogging |

Since drainage materials will always be buried in the soil, weathering is not regarded as an issue provided that the index tests and time limits on exposure described in chapter 3.3 are fulfilled. The procedure of section 1.5.1 must be followed to determine the reduction factor for weathering on the geotextiles, or to limit the exposure time.

Mechanical damage will have only an indirect effect on flow rate, but correct installation procedures are essential as with all geosynthetics. The procedure of chapter 3.6 must be followed to determine the effect of installation damage on the geotextiles.

The reduction factors, $RF_{IMCO}$ and $RF_{IMIN}$ can be set equal to 1.0 provided that original method used to measure transmissivity uses the same level of compression and type of bedding as are anticipated in the soil. While this is by no means always the case, these tests can easily be performed in the short term.

$RF_{CR}$ and $RF_{IN}$ are measured as described in 3.5.

The methods for accelerated chemical degradation described in 3.1 are not necessarily suitable for determining $RF_{CH}$ for change in flow rate after exposure, not least because of the size of sample required. In some cases, more particularly for oxidation, degradation takes place over a relatively short time following a long incubation period (see Mode 3 in 2.3 and chapter 3.1). In this case $RF_{CH}$ will remain close to 1.0 during the incubation period and then fall rapidly to zero. It may be better to make a separate calculation of the time to failure instead of applying a reduction factor (see section 1.4.4). This time to failure, modified by an appropriate safety factor, can then be compared with the design life (ref. section 3.7.2.5).

Biological degradation ($RF_{BC}$) can be caused by clogging and by root penetration for which tests such as EN 14416 exist, though not related to reduction in flow.

For drainage geosynthetics with a geonet core Giroud et al. proposed ranges of values for $RF_{CC}$ and $RF_{BC}$, together with similar ranges for $RF_{CR}$ and $RF_{IN}$.

$RF_{PC}$ depends on the soil and quality of the filter. $RF_{PC}$ can be assumed to be 1.0 as long as clogging is taken care of during design by the selection of the adjacent soil / fill in relation to the opening size of the filter.

$RF_{CD}$ is taken into account in the separate assessment of the stability of the core in relation to the long term discharge capacity. Refer to chapter 3.1, 3.5 and 3.7

Default factors are listed in section 1.5.6.

The long term discharge capacity of a drainage mat can be calculated with the short term discharge capacity value multiplied with the reduction factors

$$Q_{LT} = Q_{index} / (RF_{we} \times RF_{CR} \times RF_{IN} \times RF_{CC} \times RF_{BC} \times RF_{ID})$$

Note that separate calculations must be made for time to collapse of the core due to creep or chemical degradation in accordance with 3.7.2.5 and 3.7.7.2.

For filtration the main property is permeability, and in case of separation, the main property for design is puncture resistance. Use of the method to reduce the value of the main property of the geosynthetic with reduction factors is limited by general lack of data, however an attempt to apply the methods is made in the worked examples (section 1.5.11, case 3, 4 and 5).

For filtration we could calculate with:

$$k_{LT} = k_{index} / (RF_{we} \times RF_{CR} \times RF_{IN} \times \times RF_{CC} \times RF_{BC} \times RF_{ID})$$

## 1.4.4   Limited lifetime

When degradation is by oxidation the strength of the geosynthetic may diminish rapidly following a long incubation time. In this case it is more appropriate to predict service life than to calculate a reduction factor.

This projected service life must be more than the required lifetime, otherwise the geosynthetic is not acceptable to be installed in the project. (ref. 3.7.2.5).

When the maximum service life of a geosynthetic is critically close to the required lifetime, special quality control tests must be executed during supply and installation of the chosen product.

## 1.5   Durability assessment

Annex "B" to the application standards (EN 13249 – 13257, EN 13265 and EN 15381) describes the required results of specified tests on geosynthetics for the various durability aspects. The required results are related to a minimum predicted lifetime category, as described in section 1.5.2 and 1.5.3.

The system used in these 'index' tests is a follows:
- The tests cover polyester, polyethylene, polypropylene, polyamide 6 and polyamide 6.6 only. These however account for >95% of geosynthetics sold. Extension to cover polyvinyl alcohol (PVA) and aramid fibres is under discussion.
- The tests refer to a 'mild' environment: natural soil with 4<pH<9 and a mean soil temperature less than 25°C. These conditions apply to the large majority of applications in Europe. Contaminated or industrial soils are so varied and potentially aggressive that it would not be possible to reduce testing to simple index tests.
- Tests are specified for durations of 5 and 25 years. There is simply insufficient knowledge to define with confidence index tests for longer (e.g. 100 years) and which would apply to the wide range of geosynthetics on the market.
- The tests are intended to assure a minimum lifetime. They are not intended for precise lifetime prediction. It is likely that most geosynthetics which pass the tests will last for much longer than 25 years; on the other hand the tests should exclude materials whose durability is questionable.
- The conditions of these tests are generally extreme, necessitated by the short durations required. They are not representative of actual conditions.
- The polymers should not contain post consumer recycled material. Re-worked polymer may be used under certain conditions. In the revised EN application standards, which are expected to become valid in 2013, it is expected to be allowed to use Post Consumer Material and Post Industrial Material for service lives up to 5 years and in non-reinforcing applications.
- In general, reduction in tensile strength is the property used to determine loss of function, because it can be measured simply and reproducibly. While strength may only be indi-

rectly relevant to functions such as filtration and separation, properties such as characteristic opening size would not necessarily identify degradation while direct measurement of these functions would require the use of standardised soils and complex equipment
- For design lives extending to 100 years and other applications falling outside the above conditions a more detailed durability assessment is required. That is the subject of this document and that assessment must be executed by an expert specialized in this field (ref. 1.5.4).

In all cases evaluation is done with regards to weathering.

## 1.5.1   Weathering

Geosynthetics for all applications have to be tested for resistance to weathering. Most geosynthetics contain carbon black to increase their UV resistance, but in general they are not suitable for permanent exposure to light. EN 12224 describes accelerated weathering tests. The method of assessment differs from other durability tests in that no product fails, but the result of the test determines the maximum length of time the product may be exposed to light on site, as shown in the following table:

Table 1.5 Maximum exposure time due to weathering.

| Retained strength after testing to EN 12224 | Maximum exposure time (uncovered) during installation |
|---|---|
| >80 % | 1 month |
| 60 %-80 % | 2 weeks |
| <60 % | 1 day |
| Untested material | 1 day |

Note that this table may be changed in the revision of the EN application standards (EN 13249 – EN 13257, EN 13265 and EN 15381), which is now under discussion. It is suggested to use this table for reinforcement applications or other applications where long-term strength is required. Another table with lower requirements on retained strength is suggested to be used for another applications. As this report is focused on long term applications, we use the table as given in table 1.5.

In case the geosynthetic remains uncovered, which is for example the case with canal liners or geomembranes; special tests must be executed and evaluated. There is no method available to simulate long-term weathering within a reasonable time. This is discussed in chapter 3.3.

## 1.5.2   Required service life up to 5 years

According to the EN application standards (EN 13249 – EN 13257, EN 13265 and EN 15381) the product may be used for a service life of 5 years without further testing. The only exceptions are reinforcements or applications where long-term strength is a significant parameter

## 1.5.3  Required service life up to 25 years

For a duration of 5 years only the weathering test is required. For a duration of 25 years the material should pass the relevant screening test, with the requirement of minimum percentage retained strength of 50%:

- Polyester must be screened with EN 12447 for hydrolysis.
- Polypropylene must be screened with EN ISO 13438 for oxidation.
- Polyethylene must be screened with EN ISO 13438 for oxidation.
- Polyamide must be screened with EN 12447 for hydrolysis and with EN ISO 13438 (at 100°C) for oxidation.
- For polyvinyl alcohol a method based on EN ISO 13438 under different conditions is proposed for screening for oxidation.

A flow sheet of the various decisions to be taken to assess the durability of geosynthetics in accordance with the latest revision of the standards (EN 13249 – EN 13257, EN 13265 and EN 15381) is given in figure 1.1 a and b.

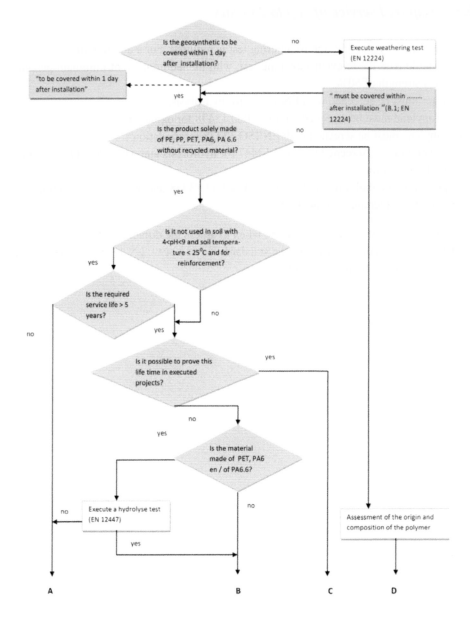

Fig. 1.1a Decisions to be taken to assess the durability of geosynthetics in accordance with (basis EN NEN 13253 e.v.) part 1.

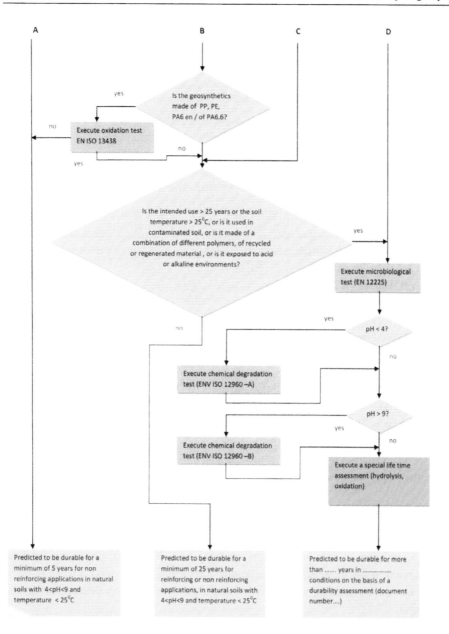

Fig. 1.1b Decisions to be taken to assess the durability of geosynthetics in accordance with (basis EN NEN 13253 e.v.) part 2.

## 1.5.4    Assessment of long term durability

An assessment of the long term durability of geosynthetics must be made in the following cases:
- The intended service life of the geosynthetic in the structure is > 25 years
- Polyester geosynthetics are used in highly alkaline environments with pH> 10, near concrete, lime or cement
- For long service lives if pH<4 or >9
- The soil temperature is more than 25°C or less than 0°C, for a longer period
- Recycled materials are used (only where a constant quality is assured).

According to the guideline for such an assessment, given in ISO/TS 13434: 2008(E), various types of degradation must be considered. Not all the mechanisms are of equal importance and they depend on the polymer used, type of application or specific conditions of use. Therefore the types of degradation must be prioritized.
- Mechanical damage (ref. chapter 3.6)
- Oxidation (ref. chapter 3.1)
- Photo-oxidation due to UV (ref. chapter 3.3)
- Hydrolysis (ref. chapter 3.2)
- Alkaline attack (ref. chapter 3.1)
- Acid attack (ref. chapter 3.1)
- Effects of solvents which may swell the polymers (ND)
- Leaching out of additives (ref. chapter 3.1)
- Stress cracking possibilities (ref. chapter 3.1)
- Effects of waste and leachates (ND)
- Compressive or tensile creep (ref. chapter 3.5 and 3.4)
- Freeze-thawing, wet-dry cycles or ion exchange (ND).

The references are made to chapters 3.1 to 3.6 in this report, where the mechanism is described in full detail together with an overview of present knowhow.
Reference (ND) means: not described as it falls outside the scope of this report. For example the effect of solvents, environmental stress cracking and, in general, waste and leachates are only relevant to thermoplastic geomembranes (geosynthetic barriers). Freeze-thawing, wet-dry cycles or ion exchange are effects specific to geosynthetic clay liners (clay geosynthetic barriers).

The assessment of durability can take the form of:
- **Reduction factors** to be applied to the key property to give a reduced value that can be used in designDefinition of a **time-dependent property** for use in design.

**Statement of lifetime**, for a single form of degradation.Use of an **index test** to establish durability for a single form of degradation.

Examples of **reduction factors** are:
- Mechanical damage (ref. 3.6)
- Compressive or tensile creep (ref. 3.5 or 3.4)
- Chapter 3.7 gives an extensive description of the back ground of the methods used to determine the reduction factors.

Some **time-dependent properties** are expressed in the form of a graph or table for use in design rather than as reduction factors. An example is tensile strain of a reinforcement.

Examples of **index tests** leading to appropriate **materials selection** include:
- Use of a robustness classification for mechanical damage (see section 3.6.3)
- Index tests, if acceptable, for the hydrolysis of polyesters (see chapter 3.2)
- Abrasion (see section 3.6.8)
- Protection efficiency (see section 3.6.9).

Section 3.8.1 gives a detailed procedure including all the steps that are to be taken to make a proper durability assessment resulting in a **statement of lifetime**. That section should be taken as a guide either to make such an assessment or to evaluate an assessment made by others. See also ISO/TS 13434, Geosynthetics – guidelines for the assessment of durability.

The most common form of degradation for which such assessments are necessary is chemical attack, as summarised in the following paragraphs.

*Oxidation*
The index test to measure the resistance to oxidation is ISO 13438. It includes two alternative test methods. In the oven tests only the temperature is increased, with method A for polypropylene and B for polyethylene (ref. 3.1.4.2). In the pressurized oxygen test (Method C), the material is placed in an aqueous solution. It is heated under 5 MPa (50 bar) of oxygen. Thus oxidation is accelerated both by temperature and by increasing the concentration of oxygen (ref. 3.1.4.4 and figure 3.1.8).

These methods for polyethylene and polypropylene can be used as the basis of Arrhenius tests to predict durability over a 100 year lifetime, or to derive the reduction factor $RF_{CH}$.

Simple oven ageing tests using the method of ISO 13438 Methods A2 or B2 can be extended over a range of temperatures and times with a defined retained strength as the criterion. In this case a chart of the rate of reduction in strength against time will be required to predict the retained strength for the design life and service temperature as in figures 3.7.4 and 3.7.5.

The pressurized oxygen test using ISO 13438 Method C1 or C2 can be applied using a range of temperatures and pressures. The reduction in strength following exposure to these conditions will be charted on a three dimensional diagram which will then be extrapolated to the service conditions to yield a predicted retained strength,

Alternatively, the degradation of stabilized polymers can be divided into three stages:
- The anti-oxidant depletion time
- The induction time to onset of polymer degradation
- The time to reach a drop of 50% of a mechanical property
  (ref. 3.1.4.3 and figure 3.6).

Simple oven ageing tests are performed as in ISO 13438 Methods A2 or B2 over a range of times and temperatures, but in addition to the strength the reduction in stabilizer content is measured by chemical analysis, OIT or HP-OIT. In the first stage the OIT is unchanged. In the second stage the OIT reduces progressively as the anti-oxidant is consumed, but the mechanical properties remain unchanged. In the third stage the reduction in strength or elongation must be measured. The tests are executed at various temperatures and Arrhenius curves are made to extrapolate the duration of each stage to lower temperatures. In this case the design life must take into account the durations of the first two stages and the retained strength calculated for the time remaining. It may be sufficient to predict the time to the end of the second stage: if this exceeds the design life then no reduction in strength is predicted (see 1.4.4).

All three methods are designed to predict a retained strength for a defined design life and environment. Then for reinforcements:

$RF_{CH}$ = initial strength/predicted retained strength.

The reference temperature for the presentation of the durability is 25° C. Normally geosynthetics are buried in soil where for European conditions 15° C would prevail (with maybe an exception for southern Europe). Temperature accelerated testing gives an opportunity of extrapolating to a lower temperature as the reference temperature. A difference of 10° C has a large impact on the life assessment!

Chapter 3.1 gives an extensive description of the oxidation mechanism, the various test methods and the evaluation of the results.

*Internal hydrolysis*
EN 12447 is an index test to assess PET geosynthetics with an expected service lifetime of 25 years. PET geosynthetics are susceptible to hydrolysis. This can be predicted by exposing the yarns to hot water and establishing an Arrhenius relation between time to a specific reduction in strength and temperature.

EN 12447 can be extended to lower temperatures and longer times to yield an Arrhenius diagram (Schmidt et al. 1994). In that way it can also be used for the prediction of longer lifetimes. See also Hsuan et al. (2008).

Chapter 3.2 describes the hydrolysis process in great detail, including the test methods and methods of interpretation of the results.

If the yarns are coated in the final product, it is adviseable to test the polyester yarns uncoated. This can only be done on virgin, uncoated material as the coating cannot properly be removed.

It is also important to notice that in soil that is not fully saturated the rate of hydrolysis decreases approximately in proportion to the relative humidity.

Note that ISO/TR 20432 allows for an index test for 100 year lifetime.

*Acid and alkaline testing*
The methods described in EN 14030 for determining the acid and alkaline attack on geosynthetics can similarly be extended to lower temperatures and longer times to yield an Arrhenius diagram (ref. 3.2.5).

*Microbiological testing*
The high molecular weight synthetic polymers commonly used in geosynthetics are in general not affected by the action of fungi and bacteria. It is not required to test virgin (not recycled) polyethylene, polypropylene, polyester (polyethylene terephthalate: PET) and polyamides 6 and 6.6. EN 12225 describes a test method which may be applied to determine whether a statistically significant loss of properties takes place over the duration of the test under conditions of maximum biological activity. It must be applied to other materials including vegetable-based products, new materials, geocomposites, coated materials and any which are of doubtful quality. Because of its limited degree of acceleration, however, the method is not a demonstration of very long-term resistance to biological degradation.

## 1.5.5  *Statement of the durability / Assessment certificate*

The durability assessment must clearly state:
* The type of material, the chemical nature and other components like additives and coatings. As producers do not like to give away the composition of compound used for their product, it may be kept confidential, but it must have been given to the certifying body and registered by them.
* The environment for which the prediction is made:
  * soil gradation, angularity, compaction
  * soil pH, presence of contaminants
  * temperature
  * saturation
  * exposure to light
* The property measured, and the end-of-life criterion
* The predicted lifetime
* The level of confidence
* Any conditions applying to the prediction, e.g. correct installation practice.

The consequences of such a prediction of lifetime can be:
* The geosynthetic is sufficiently durable:
  * No change in the available property is predicted at the end of the design life.

- • A change in the available property is predicted at the end of the design life and the level is acceptable.
- • The ratio of the predicted available property to the predicted required property at the design life is acceptable.
  - • The margin between the design life and the predicted end of life is acceptable.
- • The *geosynthetic should be replaced* after a stated number of years.
- • A *sample of geosynthetic should be extracted after a stated number of years* to determine the level of degradation; a decision regarding replacement will depend on the result.
- • If replacement or extraction is impractical, the *geosynthetic is not sufficiently durable* for the application.

## *1.5.6 Default values for the various RF*

These default values can be used in case no specific test data are known. It will be clear that the use of default values will lead to a low values in the long term properties, therefore it is advised to use as much as possible measured data for a specific product.

*Reinforcement*
$RF_{CR}$ reduction factor for creep-rupture.
In the absence of specific data an overall default factor of $RF_{CR} = 2.0$ for polyester and $RF_{CR}$ = 4.0 for polypropylene reinforcements have been proposed (Greenwood and Shen 1994, Koerner 1994). The summary of international recommendations compiled by Zornberg and Leshchinsky in 2001 gave the following default factors:

Table 1.6 $RF_{CR}$ reduction factor for creep-rupture- default value (ref. chapter 3.7.2.2).

| | |
|---|---|
| Polyester | 2.5 |
| Polypropylene | 5.0 |
| Polyethylene | 5.0 |
| Aramid | 2.5 |
| Polyamide | 2.5 |

$RF_{ID}$ : *reduction factor for installation damage*
Some tables have been issued of general values of RFID based on a comprehensive series of tests for example Hufenus et al. (2005). The user is advised to refer to the original documents to confirm that the materials and soils used were sufficiently comparable for the materials for which a life prediction is to be made.

Table 1.7 $RF_{ID}$ Reduction factors for installation damage, highest default value based on Hufenus et al. (2002) (ref. chapter 3.7.2.3).

| Geosynthetic | Fine-grained soil | Rounded coarse-grained soil | Angular coarse-grained soil |
|---|---|---|---|
| Uniaxial HDPE grids | 1.1 | 1.2 | 1.4 |
| Biaxial PP grids | 1.2 | 1.3 | 1.5 |
| PET flat rib grids | 1.1 | 1.1 | 1.1 |
| Coated PET grids | 1.1 | 1.2 | 1.3 |
| PP & PET wovens | 1.2 | 1.4 | 11.5 |
| PP & PET nonwovens | 1.1 | 1.4 | 1.5 |
| PP slit tape wovens | 1.2 | 1.3 | 1.4 |

Recently Bathurst et al. (2011a) have performed a detailed reliability analysis based on a wide range of reported values of $RF_{ID}$. The results are shown in table 3.10 (chapter 3.7). The upper limits are given here as highest default values.

Table 1.8 $RF_{ID}$ Reduction factors for installation damage, upper limit given as default value based on Bathurst et al (2011a) (ref. chapter 3.7.2.3).

| | $RF_{ID}$ for $d_{50} < 19$ mm | $RF_{ID}$ for $d_{50} > 19$ mm |
|---|---|---|
| HDPE uniaxial geogrids | 1.17 | 1.43 |
| PP biaxial geogrids | 1.11 | 1.45 |
| PVC coated PET geogrids | 1.39 | 1.85 |
| Acrylic and PP coated PET geogrids | 1.37 | 2.02 |
| Woven geotextiles | 1.66 | 4.93 |
| Nonwoven geotextiles | 1.46 | 4.96 |

When a design is made based on one brand of product normally the factors for that product should be used. If at the tender stage the contractor proposes to use a different (normally) cheaper product these factors need to be checked thoroughly before that product is approved. One might even go so far that the structure needs to be redesigned using the proper factors for the alternative product.

$RF_{WE}$ : reduction factor for weathering

The only estimates of $RF_{WE}$, the reduction factor for weathering, are those in Annex B common to EN 13249-13257 and 13265, table 3.10:

Table 1.9 $RF_{WE}$ : reduction factor for weathering – default values (ref. table 3.11).

| Retained strength after testing to EN 12224 | Time allowed for exposure on site | Reduction factor $RF_{WE}$ |
|---|---|---|
| >80% | 1 month | 1/percentage retained strength |
| 60% to 80% | 2 weeks | 1.25 |
| <60% | 1 day | 1.0 |
| Untested material | 1 day | 1.0 |

$RF_{CH}$ : *reduction factor for the environment, including chemical and biological attack*

Polypropylene and Polyethylene:

ISO TR 20432 sets out default factors for polypropylene and polyethylene, namely that if the material passes ISO 13438 Method A2 for polypropylene or B2 for polyethylene, or C2 for either, then $RF_{CH} = 1.3$ for a design lifetime of 100 years at a temperature of up to 25 °C. This default factor should be treated with great caution and is not recommended to be used (ref. section 3.7.2.7).

The reduction factor $RF_{CH}$ for polypropylene and polyethylene should preferably be based on tests which predict the reduction in strength.

Polyester:

ISO TR 20432 introduces simplified procedures for polyester geosynthetics which satisfy certain basic criteria, assuming no post-consumer or post-industrial recycled material is used. For polyester either:
• The polyester geosynthetics used for reinforcement, or the yarns from which they are made, should exhibit no more than a 50 % reduction in strength when subjected to EN 12447.

or:
• The CEG measured according to GRI-GG7 should be less than 30 meq/g, and the number averaged molecular weight, $M_n$, determined according to GRI-GG8 should be 25,000 or more.

For a geosynthetic that satisfies either recommendation used in saturated soil, $RF_{CH}$ is given in table 1.10.

Table 1.10 Default values of $RF_{CH}$ for polyester reinforcements from ISO TR 20432 at 25 °C (ref. table 3.12).

| pH range | Design lifetime (years) | Service temperature (°C) | $RF_{CH}$ |
|----------|-------------------------|--------------------------|-----------|
| 4 - 9 | 25 | 25 | 1.0 |
| 4 - 8 | 100 | 25 | 1.2 |
| 8 - 9 | 100 | 25 | 1.3 |
| 4 - 9 | 25 | 35 | 1.4 |

*Calculation of $f_s$*

The factor of safety $f_s$ included in the calculation of design strength in ISO/TR 20432 accounts solely for the following:

- uncertainty of extrapolation of creep-rupture data ($R_1$)
- uncertainty of extrapolation of Arrhenius data ($R_2$).

The minimum value for $f_s = 1.2$.

Further details are given in section 3.7.3.

*RF for reinforcements*

The total RF is calculated as follows:

$$RF = RF_{CR} \times RF_{ID} \times RF_{W} \times RF_{CH} \times f_s.$$

Note if there is a restriction on strain a separate calculation must be performed as described in 3.7.6.

*Drainage and Filtration applications*

As default values can be used, in case the transmissivity and permeability tests have been executed for a duration of 100 hrs and as long as clogging is taken care of during design by the selection of the adjacent soil / fill in relation to the opening size of the filter (section 1.4.3):

| $RF_{PC}$ | Reduction factor for particulate clogging | = 1.0 |
|-----------|-------------------------------------------|-------|
| $RF_{IMCO}$ | Reduction factor for instantaneous compression | = 1.0 |
| $RF_{IMIN}$ | Reduction factor for instantaneous intrusion by the soil | = 1.0 |

Accelerated tests must be executed and an assessment must be made of the time to failure due to chemical degradation instead of using a default $RF_{CH}$ (ref. section 3.7.2.5).

Default values for the following factors are given in table 1.11

| $RF_{CR}$ | Reduction factor for time-dependent compression of the core (compressive creep) |
|---|---|
| $RF_{IN}$ | Reduction factor for time-dependent intrusion of the soil |
| $RF_{CC}$ | Reduction factor for chemical clogging |
| $RF_{BC}$ | Reduction factor for biological clogging |

$RF_{ID}$ can be taken from table 1.7 or 1.8.

Table 1.11 Reduction factors for drainage geosynthetics with a geonet core after Giroud et al. (2000) (ref. chapter 3.7.7.3).

| Examples of application | Normal stress | Liquid | $RF_{IN}$ | $RF_{CR}$ | $RF_{CC}$ | $RF_{BC}$ |
|---|---|---|---|---|---|---|
| Landfill cover drainage layer Low retaining wall drainage | Low | Water | 1.2 | 1.4 | 1.2 | 1.5 |
| Embankment, dams, landslide repair<br>High retaining wall drainage | High | Water | 1.2 | 2.0 | 1.2 | 1.5 |
| Landfill leachate collection layer Landfill leakage collection and detection layer Leachate pond leakage collection and detection layer | High | Leachate | 1.2 | 2.0 | 2.0 | 2.0 |

For drainage mats, the total RF is:

$$RF_{tot} = RF_{we} \times RF_{CR} \times RF_{IN} \times RF_{CC} \times RF_{BC} \times RF_{ID}$$

Note that a separate calculation must be made for time to collapse of the core due to creep or chemical degradation in accordance with 3.7.7.2.

For filtration we calculate with:

$$RF_{tot} = RF_{we} \times RF_{CR} \times RF_{IN} \times \times RF_{CC} \times RF_{BC} \times RF_{ID}$$

## 1.5.7   Expert institutes

It will be clear that a durability assessment can only be done by experienced institutes or sworn experts. In Europe only a few institutes have experience to execute the tests and to do a proper assessment. A list of these institutes in the Netherlands, Germany and some in the UK is given in attachment D.

The assessment of durability for more than 25 years may in Europe only be executed by an institute certified under the CE mandate. A list of these institutes can also be found in attachment D.

## 1.5.8 *Approval procedure of geosynthetics during tendering*

The various steps in the approval procedure of geosynthetics are given in the flow scheme 1.2.

The first phase is the selection of the product, based on price, technical specifications etc. Sometimes this is done in two steps when products do not pass the requirements in the first round of evaluation.

In case a lifetime assessment and or mechanical damage factors are available for the product, no further steps have to be taken.

But in the case that one or more of the products which are under consideration do not have a lifetime assessment or reduction factors established by accredited institutes, this assessment must be done during the tendering phase.

It must be noted that the testing and assessment of the service life of a geosynthetics can take several months and is costly. Therefore it can only be done during the tendering phase in cases where large quantities of geosynthetics are required, or if there is sufficient time before the product is to be supplied.

Since the anti-oxidants added to polyolefin materials (PP, PE) are very costly, the quantities used will normally only be intended to assure a 25 year lifetime as required for CE-marking. This means that for most standard products no assessment for a service life above 25 years has been made.

In case a service life above 25 years is required, the proposed product must either have been assessed earlier or a new assessment must be made. There is no other way to know the service life of a product.

When a product is assessed by an expert and when the service life fulfils the requirements, a product can be approved.

The products with projected service lives of more than 25 years are mostly produced to order.

During production it must be checked if the required anti-oxidants are used. This can be done in a way described in paragraph 1.5.9 point 2. If this quality control during production is not possible, the checks must be made after production or after delivery at site. In that case the procedure as described in 1.5.9 point 1 must be followed.

Polyester products are not checked for anti-oxidants, as degradation is principally by hydrolysis. In this case a check on the number of carboxyl end groups and the molecular weight of the fibres is sufficient. (Ref. chapter 1.5.6). Otherwise hydrolysis tests are required.

Most commercially available polyester products have already been tested and assessed for a service life of 100 years because they are mainly used in reinforcement applications, where this lifetime requirement is an essential part of the design.

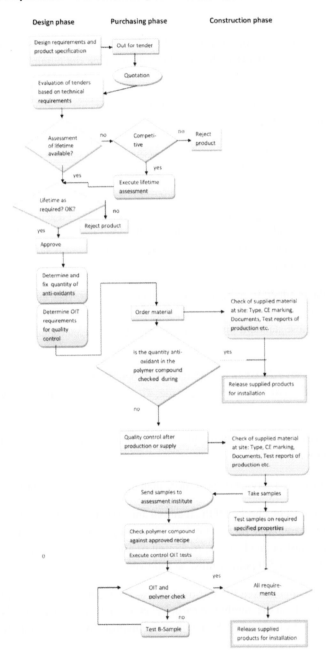

Fig. 1.2 Approval procedure of geosynthetics during tendering.

## *1.5.9    Verification of supplied material with regard to lifetime*

The various steps in quality control at site are given in attachment E (figure E.2) and shown in the flow chart figure 1.4 in chapter 1.5.8.

After delivery at site, samples must be taken and tested before the material is released for installation. All relevant properties, as indicated in the application standards (EN 13249 – EN 13257, EN 13265 and EN 15381), must be checked. According to these standards, the durability must be tested to the initial type testing of the material and this must be repeated once every 5 years or in case the polymer compound changes.

This means that in most cases, no verification of the durability aspects is done when the product is delivered at site.

However in case the service life of a geosynthetic material is of major importance for the integrity of the structure it could be advisable to execute tests on the supplied material.

Oxidation tests by means of oven tests or autoclave tests are not a realistic option because of the long execution times of the tests.

The quality control with Polyolefins PP and PE is therefore focussed on checking if the required quantity of anti-oxidant is used during production of the material. This can be done in 2 ways:

1. Fingerprinting
   The quantity and type of anti-oxidants in the polymer compound determine the oxidation time (in case of polyolefins PP, PE). During the initial type testing with exposure to high temperatures for certain times it is advisable to perform OIT tests using one of the established methods. These OIT values are representative of the initial polymer compound that has passed the durability tests. The quality is now controlled by execution of the same tests on further batches of material. The OIT is determined and compared with the OIT values measured during initial type testing at the same time as oven- or autoclave testing.
   Other chemical analysis tests can be executed which for example determine the peroxide content or the melt flow index. Depending on the polymer used, these tests can be used to fingerprint the polymer compound.

2. External inspection during production
   A more direct way is possible in case of supply of material over a longer period. In the contract documents it could be included that during the Factory Production Control audits by the CE "Notified Body" institutes, which are executed once a year, the composition of the used polymers will be checked. As the anti-oxidant additives are very expensive, it will be easier to check the incoming and outgoing material balance at the production location. As the "Notified Body" has insight in the quality control database of the producer during the normal CE- inspection and auditing, this extra information can relatively easy be checked as part of a separate agreement.

With regards to hydrolysis in case of polyester materials, it is sufficient to check the carboxyl end group count and the molecular weight distribution of the supplied products (generally fibres). As the resistance against hydrolysis does not depend on additives which are mixed into the polymer compound during production, the quality control during production is not so important. It is sufficient to check the values at product approval stage and in most cases the polyester products have already had lifetime assessments made.

### 1.5.10  Quality control according to CEN standards

The quality control of geosynthetics in Europe is regulated by various EN standards and consists of various steps:

1. Quality control by the producers during production. This is laid down in procedures which may be approved by a certification institute under ISO 9001 and ISO 9002.
2. Based on this Factory Production Control System, the producer specifies the value for various properties of the material in his product data sheets.
3. According to the CEN Attestation of Conformity system 2+, a certified "notified body" institute inspects controls and certifies this Factory Production Control System. This is declared by a Factory Production Control Certificate which is issued by the "notified body" to the Producer.
4. Before the "notified body" can issue a Factory Production Control Certificate, initial inspection of the quality control system is made, including an evaluation of the Factory Production Control System. The Factory Production Control System is audited twice a year and at that time it is checked if the declared properties of the products are in accordance with the results of the Factory Production Control System. If necessary the values of the properties are adjusted.
5. The producer must draw up a declaration of conformity in which he declares that the product is produced to the specified European Standard (e.g. EN 13249:2005)
6. Based on the Factory Production Control Certificate and the Declaration of Conformity, the producers must mark the products with a CE-Marking and deliver the products with a Accompanying Document which give the values for the harmonized characteristics, together with the tolerances for these values.

These various steps are shown in figure 1.3.

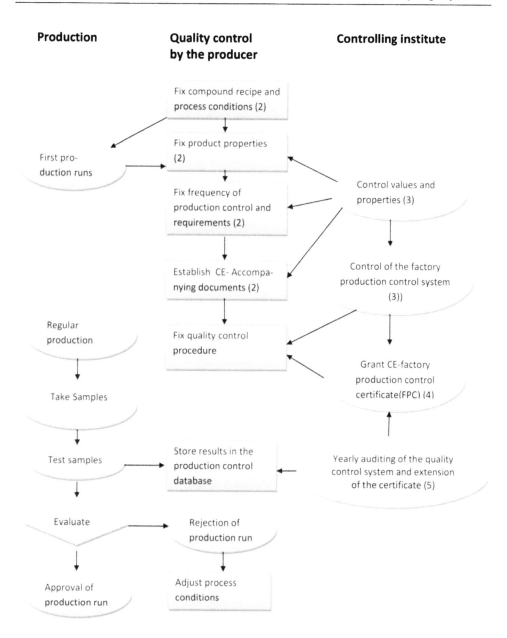

Fig. 1.3 Steps in the production quality control system under CEN regulations, the numbers between brackets refer to the points in section 1.5.10.

All steps in the quality control process to EN standards are described more detailed in attachment E.

## 1.5.11  *Worked examples*

In this section we give guidelines to calculate the long term properties of geosynthetics with RF's and to compare products.

Some products have already been assessed on durability and they have an assessment certificate. Other products are sometimes not yet assessed. They must be tested and assessed before a comparison with other products can be made or alternatively default values can be used in the calculation of the long term property.

The various steps which will be described are based on the information in section 1.4 and 1.5.

Note that in these worked examples, we do not make an assessment of the service life or the long term properties of the product. We show how the calculation of the long term properties with RF's is made and how a comparison must be made between products in case a selection or approval of a product is required for the execution of a project.

*Steps to be taken to determine the long term properties of a geosynthetic based on durability*

*Step 1:*
Determine the relevant durability mechanisms for the function of the geosynthetics in the specific application with tables 1.1 and 1.2.

Specify the main properties.

Decide which approach to take (see the box in 1.4.1, also 2.3.5. and 3.8.1)

*Step 2:*
Specify the applicable RF's based on table 1.3 for these main properties.

*Step 3:*
Determine RF's Chemical degradation (oxidation or hydrolysis)

Check the availability of Certificates which specify the service life of the product. Check also if RF's have been determined by certified institutes, based on separate tests (for example mechanical damage tests). Note that these certificates must have been issued by certified and accredited institutes or sworn experts (refer to chapter 1.5.7 and attachment D).

For products which have been marked with a CE-mark, the CE-accompanying document or the CE- Product data sheet state the expected service life (predicted to be durable for a period of at least 5 years for non reinforcing applications or alternatively predicted to be durable for a period of at least 25 years for all applications).

Some products have been certified for a service life of 100 or 120 years in special applications. Check if the application and environmental condition listed in the certificate is applicable for the considered use. Further check if the assessment method used is the same as described in this report (section 1.5.4 and 1.5.5)
Examples of such certificates can be found at:

**British Board of Agrément:**

http://www.bbacerts.co.uk/certificate_search_results.
aspx?SearchType=Product&ProductType=2355&KeywordSearch=geosynthetic

**Gov. Hong Kong CEDD:**

http://www.cedd.gov.hk/eng/services/certification/cert_pdf.htm Bundesanstalt für Materialforschung und -prüfung,

**BAM, Berlin:**

http://www.bam.de/de/service/amtl_mitteilungen/abfallrecht/abfallrecht_medien/tabelle_
kunststoffdichtungsbahnen_u_hersteller_1.pdf

Determine the $RF_{CH}$ based on chapter 3.1 or 3.2 and 3.7. See also chapter 1.5.6 In special cases default values may be applied.(see chapter 1.5.6 for details):

*Polypropylene or Polyethylene:*
ISO TR 20432 sets out default factors for polypropylene and polyethylene, namely that if the material passes ISO 13438 Method A2 for polypropylene or B2 for polyethylene, or C2 for either, then $RF_{CH} = 1.3$ for a design lifetime of 100 years at a temperature of up to 25 °C. This default factor should be treated with great caution and is not recommended to be used (ref. section 3.7.2.7).

The reduction factor $RF_{CH}$ for polypropylene and polyethylene should preferably be based on tests which predict the reduction in strength.

Note: In some cases, more particularly for oxidation, degradation takes place over a relatively short time following a long incubation period (see Mode 3 in chapter 2.3 and chapter 3.1). In this case RFCH will remain close to 1.0 during the incubation period and then fall rapidly to zero. It may be better to make a separate calculation of the time to failure instead of applying a reduction factor (see section 1.4.4). This time to failure, modified by an appropriate safety factor, can then be compared with the design life.

*Polyester:*

If the polyester geosynthetics used for reinforcement, or the yarns from which they are made, should exhibit no more than a 50 % reduction in strength when subjected to EN 12447. Or the CEG measured according to GRI-GG7 should be less than 30 meq/g, and the number averaged molecular weight, $M_n$, determined according to GRI-GG8 should be 25,000 or more.

One or the other criterion should be satisfied: for the corresponding values of $RF_{CH}$ see table 1.10.

In all other cases a special durability assessment must be made and tests executed.

Weathering
Check the availability of Certificates which specify the maximum time of uncovered use. Use this time as a requirement for installation.

Note that these certificates must be granted by certified and accredited institutes or sworn experts. (refer to chapter 1.5.6 and attachment D).

In Europe this information can be found in the CE-accompanying documents.

In case this information is not available, determine the $RF_W$ based on chapter 1.7.6, table 1.9. See also chapter 3.3 and 3.7

If no further tests are made the material must be covered within 1 day.

In case of reinforcement application
*   Tensile creep
    Determine the RF based on tests executed on the product. The supplier must present the necessary test data. Check if they are executed in accordance with the valid standards and by experienced laboratories.
    In case no test data is available, use the default values of chapter 1.5.6, table 1.6
*   Installation damage
    Determine the RF based on tests executed on the product. The supplier must present the necessary test data. Check if they are executed in accordance with the valid standards and by experienced laboratories.
    In case no test data is available, use the default values of chapter 1.5.6, table 1.7 and 1.8
*   Determine $f_s$
    Determine $f_s$ using the values $R_1$ and $R_2$.
    The value of fs must be determined using the equation:
    $f_s = 1 + \sqrt{\{(R_1 - 1)^2 + (R_2 - 1)^2\}}$.
    In case default values are used $f_s = 1,2$.(ref. chapter 1.5.6).

In case of drainage and filtration applications
*   Compression creep (if applicable)
    Check the availability of Certificates which specify the service life of the product and

or the RF's to be used. Note that these certificates must be granted by certified and accredited institutes or sworn experts. (refer to chapter 1.5.7 and attachment D). Some drainage mats have been certified for 100 years service life by BAM.

No default values can be used to replace long term testing. Different types and thicknesses of the core can result in large differences in behavior of the core, even leading to sudden collapse, therefore oven ageing tests coupled with determination of the depletion of antioxidants are recommended as the most reliable measurement. The same applies for evaluation of the filter mat in case of application in a filter structure. Accelerated tests must be executed and an assessment of the lifetime of the product must be made before an approval can be given.

- Installation damage
  Determine the RF based on tests executed on the product. The supplier must present the necessary test data. Check if they are executed in accordance with the valid standards and by experienced laboratories.

In case no test data is available, use the default values of chapter 1.5.6, table 1.7 and 1.8.

Step 4: Apply the RF"s to the main property of the geosynthetic
Multiply the RF's as described in chapter 1.5.6.
Divide the index values of the relevant properties by this $RF_{tot}$ to get the Long Term Design Value (LTDV) of the properties.
This LTDV must be more than the required design value of the relevant property.

Worked examples
In these examples a comparison is made between a product for which durability data is available and one which does not have durability data and for which default values must be used.

Preferably only products which have been assessed should be used for long lifetimes. Only in these cases can valid comparisons be made with other products.

## Case 1. Reinforced embankment

Application: reinforcement

Required service life 120 years

PE geogrid

Fill material maximum particle size 37.5 mm

Soil pH = 6

| Product A, | with CE-mark and BBA certificate |
| | char. tensile strength $T_{char}$ = 57 kN/m |
| Product B, | with CE mark, no long term durability assessment |
| | char. tensile strength $T_{char}$ = 57 kN/m |

Both products pass the screening test of ISO 13438.

Step 1: relevant durability mechanisms:

| Oxidation (PP, PE) | $RF_{CH}$ |
|---|---|
| Hydrolysis (PET) | Not applicable as the product is not made of PET |
| Weathering | $RF_{WE}$ |
| Tensile creep | $RF_{CR}$ |
| Installation damage | $RF_{ID}$ |

Step 2:
Main property: Tensile strength

$$RF_{tot} = RF_{CR} \times RF_{ID} \times RF_{WE} \times RF_{CH} \times f_s$$

Step 3:

| | product A | value from: | **Product B** | value from: |
|---|---|---|---|---|
| $RF_{CH}$ | 1,00 | BBA | 1,30 | chapt. 1.5.6[*) ] |
| $RF_{WE}$ | 1,00 | BBA | 1,00 | CE-Acc. Doc |
| $RF_{CR}$ | 2,38 | BBA | 5,00 | table 1.6 |
| $RF_{ID}$ | 1,07 | BBA | 1,17 | table 1.8 |
| fs | 1,00 | BBA | 1,20 | chapt. 1.5.6 |
| | | | | |
| $RF_{tot}$ | 2,54 | | 9.13 | |

[*)] See note in section 1.5.11 under *Polypropylene or Polyethylene* : in case of doubt, it may be better to make a separate calculation of the time to failure instead of applying a reduction factor.

Step 4:
Long term design strength: $T_D = T_{char} / RF_{tot}$
Product A: $T_D = 57 / 2,54 = 22,44$ kN/m

Product B: $T_D = 57 / 9.13 = 6.24$ kN/m.
Note that although both products have the same index test strength $T_{char}$, the long term design strength (allowable design strength) varies a lot as result of the use of the default values.

Both products must be covered within 1 month.

## Case 2. Reinforced embankment

Application: basal reinforcement

Required service life 60 years

Expected in soil temperature: 15 °C

PET flat rib geogrid

Fill material maximum particle size 37.5 mm

Soil pH = 6

Product A, with CE-mark and BBA certificate, char. tensile strength $T_{char}$ = 190 kN/m

Product B, with CE mark, no long term durability assessment, char. tensile strength $T_{char}$ = 190 kN/m

Both products pass the screening of ISO 13438.

Step 1:relevant durability mechanisms:

| Oxidation (PP, PE) | Not applicable as the product is not made of PP or PE |
|---|---|
| Hydrolysis (PET) | $RF_{CH}$ |
| Weathering | $RF_{WE}$ |
| Tensile creep | $RF_{CR}$ |
| Installation damage | $RF_{ID}$ |

Step 2:

Main property: Tensile strength

$$RF_{tot} = RF_{CR} \times RF_{ID} \times RF_{WE} \times RF_{CH} \times f_s$$

Step 3:

| | product A | value from: | Product B | value from: |
|---|---|---|---|---|
| $RF_{CH}$ | 1,00 | BBA | 1,20 | chapt. 1.5.6 |
| $RF_{WE}$ | 1,5 | BBA | 1,00 | CE-Acc. Doc |
| $RF_{CR}$ | 1,00 | BBA | 2.5 | table 1.6 |
| $RF_{ID}$ | 1,00 | BBA | 1,20 | table 1.7 |
| fs | 1,05 | BBA | 1,20 | chapt. 1.5.6 |
| | | | | |
| **RFtot** | 1.58 | | 3.96 | |

Step 4:

Long term design strength: $T_D = T_{char} / RF_{tot}$

Product A: $T_D = 190 / 1.58 = 120.25$ kN/m

Product B: $T_D = 190 / 3.96 = 47.97$ kN/m.

Note that although both products have the same index test strength $T_{char}$, the long term design strength (allowable design strength) varies a lot as result of the use of the default values.

Both products must be covered within 1 month.

### Case 3. Drainage mat

Application: drainage

Required service life 100 years

PP drainage mat

Fill material maximum sand

Soil pH = 6

| Product A, | with CE-mark and BBA certificate |
|---|---|
| | flow capacity in plane q = 0.9 1/m.s |
| Product B, | with CE mark, no long term durability assessment |
| | flow capacity in plane q = 0.9 1/m.s |

Both products pass the screening test of ISO 13438.

Step 1: relevant durability mechanisms:

| Oxidation (PP, PE) | $RF_{CH}$ |
|---|---|
| Hydrolysis (PET) | Not applicable as products are made of PP |
| Weathering | $RF_{WE}$ |
| Compression creep | $RF_{CC}$ |
| Composed of detailed values for $RF_{CR}$, $RF_{IN}$, $RF_{CC}$, $RF_{BC}$ | |
| Installation damage | $RF_{ID}$ |

With according to chapter 1.4.3:

| $RF_{CR}$ | Reduction factor for time-dependent compression of the core (compressive creep) |
|---|---|
| $RF_{IN}$ | Reduction factor for time-dependent intrusion of the soil |
| $RF_{CC}$ | Reduction factor for chemical clogging |
| $RF_{BC}$ | Reduction factor for biological clogging |

Step 2:

Main property: flow capacity in plane

$$Q_{LT} = Q_{index} / (RF_{WE} \times RF_{CR} \times RF_{IN} \times RF_{CC} \times RF_{BC} \times RF_{ID})$$

$Q_{index}$ is a flow rate determined under simulated conditions for 100 hr. duration on the drainage mat according to the test method, described in GRI-GC8. Therefore: $RF_{IMCO}$ and $RF_{IMIN}$ are equal to 1.0. $RF_{PC}$ can be assumed 1.0 as particle clogging is taken care of during of the design.

Step 3:

| | product A | value from: | Product B | value from: |
|---|---|---|---|---|
| $RF_{CH}$ | Incl. | BAM | Assessment to be made | |
| $RF_{WE}$ | Incl. | BAM | 1,00 | CE-Acc. Doc |
| $RF_{CR}$ | Incl. | BAM | 1,40 | table 1.11 |
| $Rf_{IN}$ | Incl. | BAM | 1,20 | table 1.11 |
| $RF_{CC}$ | Incl. | BAM | 1,20 | table 1.11 |
| $RF_{BC}$ | Incl. | BAM | 1,50 | table 1.11 |
| $RF_{ID}$ | Incl. | BAM | 1,10 | table 1.7 |
| | | | | |
| $RF_{tot}$ | 3,10 | | 3.33 | |

Step 4:
The Long term flow capacity for product A is according to the BAM certificate: 0,29 l/m.s
Product B cannot be evaluated as no long term assessment is made of the effects of oxidation on the stability of the core. Therefore accelerated tests must be executed and an assessment must be made in case it will be used for a service life > 25 years. Product B is to be rejected without assessment.

Both products must be covered within 1 month.

**Case 4. Filter layer**
Application: filtration

Required service life 100 years

PP nonwoven, 300 gr/m, $O_{90}$ =0,10 mm permeability 110 mm/s

Fill material maximum sand

Soil pH = 6

| Product A, | with CE-mark and BBA certificate |
| | $O_{90}$ =0.10 mm permeability k =110 mm/s |
| Product B, | with CE mark, no long term durability assessment |
| | $O_{90}$ =0.10 mm permeability k = 110 mm/s. |

Both products pass the screening test of ISO 13438.

Step 1: relevant durability mechanisms:

| Oxidation (PP, PE) | $RF_{CD}$ |
| Hydrolysis (PET) | |
| Weathering | $RF_{WE}$ |
| Time-dependent intrusion of the soil | $RF_{IN}$ |
| Particulate clogging | $RF_{PC}$ |
| Chemical clogging | $RF_{CC}$ |
| Biological clogging | $RF_{BC}$ |
| Compression creep | $RF_{CR}$ |
| Installation damage | $RF_{ID}$ |

With according to chapter 1.5.6:

| $RF_{CR}$ | Reduction factor for compression creep |
| $RF_{IN}$ | Reduction factor for time-dependent intrusion of the soil |
| $RF_{CD}$ | Reduction factor for chemical degradation |
| $RF_{PC}$ | Reduction factor for particulate clogging |
| $RF_{CC}$ | Reduction factor for chemical clogging |
| $RF_{BC}$ | Reduction factor for biological clogging |

$k_{index}$ is a permeability determined under simulated conditions for 100 hr. duration on the drainage mat according to the test method, described in GRI-GC8. Therefore: $RF_{IMCO}$ and $RF_{IMIN}$ are equal to 1.0. $RF_{PC}$ can be assumed 1.0 as particle clogging is taken care of during of the design.

Step 2:
Main property: permeability perpendicular to the plane

$$k_{LT} = k_{index} / (RF_{WE} \times RF_{CR} \times RF_{IN} \times \times RF_{CC} \times RF_{BC} \times RF_{ID})$$

Step 3:

| | product A | value from: | Product B | value from: |
|---|---|---|---|---|
| $RF_{CH}$ | | BAM | Assessment to be made | chapt. 1.5.6 |
| $RF_{WE}$ | | BAM | 1.00 | CE-Acc. Doc |
| $RF_{PC}$ | | BAM | 1.00 | chapt. 1.5.6 |
| $RF_{CC}$ | | BAM | 1.20 | table 1.11 |
| $RF_{BC}$ | | BAM | 1.50 | table 1.11 |
| $RF_{CR}$ | | BAM | 2.50 | See remark below |
| $RF_{ID}$ | | BAM | 1.10 | table 1.7 |
| RFtot | 2.44 | | 4.95 | |

The default values from table 1.11 are based on drainage mats with a geonet core. Only those default values which can be directly related to the nonwoven filter layers at the outside of the drainage mat can be used in this calculation ($RF_{WE}$, $RF_{IN}$, $RF_{CC}$, $RF_{BC}$ and $RF_{ID}$).
No default data is available for $RF_{CR}$. To be able to make a comparison it is advised to make compression creep testing or to use a safety value of 2,5.

Step 4:
The Long term permeability for product A is according to the BAM certificate: $k_{LT}$ 60- 15 mm/s. Product B cannot be evaluated as no long term assessment is made of the effects of oxidation on the stability of the filter layer. Therefore accelerated tests must be executed and an assessment must be made in case it will be used for a service life > 25 years.

Product B is to be rejected without assessment.

The durability aspects do not influence the $O_{90}$ values.

Both products must be covered within 1 month.

# References

*Chapter 0 – 1.3*
- Guidelines for the design and construction of flexible revetments incorporating geotextiles for inland waterways, PIANC 1987.
- Guidelines for the design and construction of flexible revetments incorporating geotextiles in marine environment, PIANC 1992.
- G. v. Santvoort e.a., Geotextiles and geomembranes in civil engineering, Balkema, Rotterdam, 1994.
- R.M. Koerner, Designing with geosynthetics, Prentice-Hall, Englewood Cliffs, NJ 07632, 1987.
- P. Giroud, Geosynthetics International, Special section on Liquid Collection Systems, Vol. 7, 2000.
- Federal Waterway Engineering and Research Institute (Bundesanstalt für Wasserbau) BAW: Code of Practice Use of Geotextile Filters on Waterways (MAG), 1993.
- CUR-rapport 174, Geokunstoffen in de Waterbouw, tweede herziene uitgave, CUR, Gouda, january 2009.
- CUR-rapport 151, Geokunststoffen in de civiele techniek. CUR, Gouda, 1991.
- CUR-rapport 217, Ontwerpen met geotextiele zandelementen, CUR, Gouda, july 2006.
- CUR-rapport 214, Geotextiele zandelementen, CUR, Gouda, 2004.
- G.v. Santvoort, Geotextiles and Geomembranes in Civil Engineering, Balkema, Rotterdam, 1994.
- PIANC MarCom 56, The Application of Geosynthetics in Waterfront Areas, Report 113, 2011.
- CUR-rapport 206, Geokunststoffen op de bouwplaats. CUR, Gouda, 2001.
- CUR-Aanbeveling 115: 2011, Uitvoering van geokunststoffen in de waterbouw.

*Chapter 1.4 - end*
- Hsuan Y G, Schroeder H F, Rowe K, Müller W, Greenwood J H, Cazzuffi D, Koerner R M.(2010), Long-term performance and lifetime prediction of geosynthetics. EuroGeo4, Edinburgh, UK, pp 327-366.
- Schmidt, H.M., Te Pas, F.W.T., Risseeuw, P. and Voskamp, W., (1994), the hydrolytic stability of PET yarns under medium alkaline conditions, Fifth International Conference on Geotextiles, Geomembranes and Related Products, Singapore, IGS, pp. 1153-1158.
- EN 12224:2000 Geotextiles and geotextile-related products - Determination of the resistance to weathering.
- EN ISO 10722:2007 Geosynthetics - Index test procedure for the evaluation of mechanical damage under repeated loading - Damage caused by granular material (ISO 10722:2007).
- EN ISO 13431:1999 Geotextiles and geotextile-related products – Determination of tensile creep and creep rupture behaviour (ISO 13431:1999).
- EN ISO 13438:2004 Geotextiles and geotextile-related products – Screening test method for determining the resistance to oxidation (ISO 13438:2004).

- EN ISO 25619-1:2008 Geosynthetics - Determination of compression behaviour - Part 1: Compressive creep properties (ISO 25619-1:2008).
- EN ISO 25619-2:2008 Geosynthetics - Determination of compression behaviour - Part 2: Determination of short-term compression behaviour (ISO 25619-2:2008).
- ISO/TR 13434:2008 Geosynthetics -- Guidelines for the assessment of durability.
- Annex A to EN ISO 13438:2004, "Extensive background information on oxidation processes and oxidation measurement".
- ISO/TR 15010:2005 Geosynthetics – Quality control at the project site.
- EN 12224:2000 – Geotextiles and geotextile-related products – Determination of the resistance to weathering.
- EN 12225:2000 – Geotextiles and geotextile-related products – Method for determining the microbiological resistance by a soil burial test.
- EN 12226:2000 – Geotextiles and geotextile-related products – General tests for evaluation following durability testing.
- EN 12447:2001 – Geotextiles and geotextile-related products – Screening test method for determining the resistance to hydrolysis in water.
- EN 14030:2001 – Geotextiles and geotextile-related products – Screening test method for determining the resistance to acid and alkaline liquids.
- EN 13249:2000: 2001, Geotextiles and geotextile-related products – Characteristics required. - for use in the construction of roads and other trafficked areas (excluding railways and asphalt inclusion).
- EN 13250:2001, Geotextiles and geotextile-related products. Characteristics required for use in the construction of railways.
- EN 13251:2001, Geotextiles and geotextile-related products. Characteristics required for use in earthworks, foundations and retaining structures.
- EN 13252:2001, Geotextiles and geotextile-related products. Characteristics required for use in drainage systems.
- EN 13253:2001, Geotextiles and geotextile-related products. Characteristics required for use in erosion control works (coastal protection, bank revetments).
- EN 13254: 2001, Geotextiles and geotextile-related products - Characteristics required for the use in the construction of reservoirs and dams.
- EN 13255: 2000, Geotextiles and geotextile-related products - Characteristics required for use in the construction of canals.
- EN 13256: 2001 Geotextiles and geotextile-related products - Characteristics required for use in the construction of tunnels and underground structures.
- EN 13257: 2001 Geotextiles and geotextile-related products. Characteristics required for use in solid waste disposals.

# CONTENTS

# CHAPTER 2
# INTRODUCTION OF GEOSYNTHETIC

According to EN ISO 10318, a geosynthetic is a product of which at least one component is made from a synthetic or natural polymer, in the form of a sheet, strip or three-dimensional structure, and which is used in contact with soil and/or other materials in geotechnical and civil engineering.

Geosynthetics improve the overall long-term stability and integrity of a soil structure, and have the ecological and economical advantages that they require less transportation and less handling of natural resources. With more than 50 years of successful experience, geosynthetics are well established in many applications in civil engineering. There is a large variety of materials and products, and the number of different uses for them is growing continuously. For more information on the history of geosynthetics, reference can be made to Santvoort (1994) and Koerner (2005). Recent applications are described in Pilarczyk (2001), Saathoff (2003), Heibaum (2006) and Kim (2009).

Features common to all geosynthetics are the thickness of the sheets, ranging from 1 mm for geomembranes and thin thermal bonded nonwovens to over 1 cm for geosynthetic clay liners and 2 cm for composites, and their roll widths of between 2 and 9 m. Some special geosynthetics such as geonets, containers and tubes are delivered in a different manner. This document describes their durability and design for long-term applications. Geotextiles made of biodegradable material such as coir are therefore excluded.

A geosynthetic is not just a trade name. For the sake of life prediction we need to know its structure, the polymer it consists of, and the properties essential to its function. Under structure is meant its general form, such as a textile or a sheet, the degree of orientation, the type of weave, the thickness or mass per unit area, the nature of the joints/welds and the coating, all of which require acquaintance with the methods of manufacture. The polymer should be defined not just in general terms (e.g. polyethylene, polyamide) – not always easy due to the number of copolymers and blends (mixtures) on the market – but also in terms of molecular weight and distribution, kind and content of copolymers and its additives, fillers, its density and production process. Creep-strength, pore size and permeability are examples of the properties which are key to the function for which the geosynthetic has been selected.

## 2.1    Defining the geosynthetic

J.H. Greenwood, W. Voskamp

### 2.1.1    Types of geosynthetic

ISO TR 13434, groups the principal types of geosynthetic as follows. PIANC report MarCom 56 (2011) gives more details, some of which quoted below. Formal definitions are given in the standard EN ISO 10318.

A **geotextile** is a planar, permeable, polymeric textile material, which may be manufactured by different processes and includes principally
- Woven materials
- Knitted materials
- Non-woven materials, either thermally bonded or needle punched.

The principal polymers used for geotextiles are polypropylene (PP) and polyester (PET) .

Fibres are produced from the molten polymer by extrusion through a die, cooled, and then stretched by further drawing. The polymer chains align themselves with the fibre, the orientation and the crystallinity increases. Thinner fibres may be more heavily drawn but have a higher surface-to-volume ratio which plays an important part in several superficial degradation processes like weathering and biodegradation. Fine fibres are also more easily damaged mechanically.

After drawing, the fibres can be cut to staple fibres which are stacked in random orientation, or left uncut and laid on a belt to form a continuous filament or spun bonded nonwoven geotextile. The nonwovens are bonded mechanically by needle-punching, stitching or sewing, by heat or by the use of adhesive. Needle-punching process is the most common method and is generally used to produce fabric weights of 150 g/m$^2$ up to 4000 g/m$^2$. The thickness and related properties increase proportionally. In the use of heat (melt bonding) the continuous filaments or staple fibres are melted together and the resultant fabrics are rather stiff in texture and feel. The method of bonding has a considerable influence on their properties. Nonwovens are characterized by opening size, mass per unit area and thickness and their pore volume, which depending on compressive pressure can be up to 90%, higher than the pore volume of soils.

Film tapes and split tape yarns are produced by extruding a film, cutting it into individual tapes, stretching, thermal fixing and fibrillation. Tapes and fibres can be woven using conventional textile machinery into a wide range of fabric weaves, comprising the warp in the machine direction and the weft in the cross machine direction. These woven fabrics are relatively thin and have a high tensile strength.

Fig. 2.1 Woven Geotextile.

Fig. 2.2 Non-woven geotextile (courtesy Ten Cate Geosynthetics).

**Knitted geotextiles** generally have preferentially aligned yarns and can be used as an alternative to woven geotextiles. The warp and weft yarns remain straight, elongate less during loading and are connected by knitting threads. In addition, a nonwoven geotextile can be inserted at the same time to create a geocomposite, offering unique filtration and reinforcement properties in one sheet.

The strengths of knitted can be very high, exceeding 1,000 kN/m, and the manufacturing process also allows the use of glass and carbon fibres as well as polyaramids. Of these only polyaramids are covered in this publication.

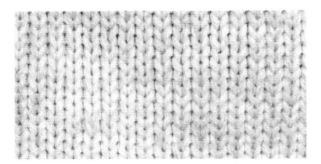

Fig. 2.3 Knitted fabric (myfreetextures.com).

A **geomembrane**, referred to in international standards as a geosynthetic barrier, is a planar, relatively impermeable to aqueous solutions, polymeric or bituminous sheet. The polymers used to manufacture the geomembranes are generally thermoplastic, elastomeric or modified bituminous materials, including high density polyethylene(HDPE), linear low density poly-ethylene(LDPE), polyvinyl chloride(PVC), flexible polypropylene(fPP), ethylene propylene diene monomer(EPDM), ethylene inter-polymer alloy(EIA), chlorinated polyethylene(CPE), chlorosulfonated polyethylene(CSPE) and other elastomers(CR et al). Geomembranes are made by extruding a sheet by a variety of different methods. On site they are bonded by thermal welding, with or without pressure, by fusion or by chemical bonding. PVC and CSPE geomembranes are manufactured by calendering methods and may consist of multiple plies with a geotextile (scrim) as support. A bituminous geomembrane is made by passing a geo-textile through consecutive baths containing modified bituminous formulations to obtain a bitumen-impregnated geotextile product.

Fig. 2.4 Smooth and textured surfaces of extruded HDPE geomembranes (wikipedia).

A **geosynthetic clay liner** (GCL) or barrier is a factory-manufactured geosynthetic hydraulic barrier consisting of clay, bentonite or other very low permeability material sandwiched between geotextiles and held together by bonding such as needle punching or stitching. The clay absorbs water from the soil, becomes hydrated and swells. Particularly relevant to its durability are e.g. the needle punching fibres, the stitch bonding filaments or yarns, the glues, the ion exchange between the barrier material and the liquid retained or contained, and desiccation and freezing-thawing resistance.

Fig. 2.5 Geosynthetic Clay Liners (GCL) (Hongyvan International Co. ltd).

A **geogrid** is a regular open semi-rigid network, whose apertures allow interlocking with the surrounding soil. Geogrids can be made from

*   Bundles of fibres, generally coated, linked by bonding or interlacing. Weaving, knitting and welding are all used in their manufacture, using fibres of polyethylene (PE), polypropylene (PP), polyester (PET), polyvinyl alcohol (PVA) and aramid (PPTA). Coating materials include acrylic polymers (ANM), polyvinyl chloride (PVC), and polyethylene (PE).
*   Laid geogrids: Laid geogrids are made from extruded strips, bars or bar shaped elements. They are laid crosswise and are flexible at the junctions (e. g. by coating) or fixed (e. g. by friction welding or laser-technique).
*   Perforated sheets of polyethylene (PE) and polypropylene (PP) which are then stretched in one or two directions. The material between the perforations is drawn stretched to produce long ribs of highly oriented polymer. Geogrids of this type are referred to for short as extruded geogrids, even though this description is not entirely correct.

The major function of geogrids is in the area of soil reinforcement as they are characterized by high tensile strength at low deformation. The key feature of all geogrids is that the apertures – the openings between adjacent longitudinal and transverse ribs –allow the geogrid to hook or anchor itself in the ground by soil-geogrid-interaction(called interlocking effect). The junctions are called nodes. Thus the mechanical and durability properties of the nodes are essential for the function of geogrids.

Fig. 2.6 Geogrids (Tensar geogrids).

A **geonet (GNT)** consists of a set of parallel ribs overlying another set, perpendicularly or diagonally, and integrally connected with it. Geonets allow water to flow laterally and are therefore mainly used as incompressible drainage layers under high compressive surcharges. They can be placed between permeable layers to allow water to drain from a wet soil, or in combination with impermeable layers, for example at the base of a landfill.

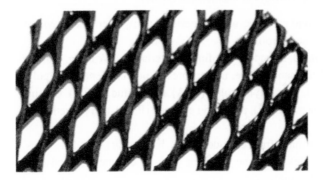

Fig. 2.7 Geonets (GSE Environmental).

A **geocell (GCE)** is a three-dimensional, permeable, polymeric (synthetic or natural) honeycomb or web structure, made from strips of geotextiles or geomembranes linked alternating, and used, for example, to hold soil particles, roots and small plants in geotechnical and civil engineering. Often called geo-cellular containment, Geocells function by transferring the downward load into the transverse plane of the soil-geocell structure.

Fig. 2.8 Geocells (WestConstruct).

A **geomat** is a three-dimensional, permeable, natural, or synthetic polymeric structure, made of bonded yarns, used to reinforce roots of grass and small plants and extend the erosion-control limits of vegetation for permanent erosion control applications. The polymers used to manufacture geomats are the same as for geosynthetics with the addition of polyamides.

Fig. 2.9 Geomat.

A **geocomposite** is a combination of two or more geosynthetics. Drainage geocomposites consist of an open structure (geospacer) typically sandwiched between two layers of nonwoven filter. Water can penetrate the nonwoven filter and then drain laterally through the spacer.

79

Fig. 2.10 Geocomposites.

A **geofoam** is a block or a planar section of rigid cellular foam polymeric material (polystyrene) commonly used as a lightweight fill to take up differential thermal expansion and for use in frozen ground, or to reduce the settlement of soft compressible soils due to the weight of fill material.

Fig. 2.11 Geofoam.

## 2.1.2   Types of polymer

Geosynthetics are made from thermoplastic polymers formed by extrusion and drawing. The polymers consist of long chain molecules made up of identical (homopolymer) or different (copolymers) monomers repeated many times over. Attached to these main polymer chains may be short side groups as well as longer chain branches. A longer chain should improve the mechanical properties but may make the polymer more difficult to process. The mechanical properties will also be influenced by the bonds within the chains and the bonds between them as well as the amount, kind and arrangement (alternating, periodic, statistical or block) of chain branching.

80

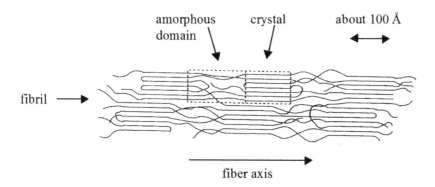

Fig. 2.12 Amorphous and crystalline domains (v.d. Heuvel, 1993).

In amorphous materials the polymer chains are arranged randomly, while semi crystalline materials contain 'crystallites' distributed within the amorphous areas, in which the chains fold or are otherwise aligned parallel to one another. The proportion of these crystallites, known as the crystallinity, will affect the mechanical and durability properties because the tightly packed molecules within the crystallites result in dense, stiff regions which liquids and gases find it hard to penetrate. 'Tie' molecules which link one crystallite with another through an amorphous region have a strong influence on the mechanical properties. Sections of the main polymeric chain containing side chains are generally located in amorphous areas. The arrangement and kind of side chains clearly influence morphology and the content of tie molecules. Thus by this instrument a design of special morphologies with tailored mechanical properties and chemical resistances is possible.

When heated above a particular glass transition temperature $T_g$ (measured by Differential Scanning Calorimetry) the amorphous regions change from being stiff, glassy and brittle to being ductile and rubbery. This change is less marked in semi-crystalline polymers, but it explains why polyesters which are used below their $T_g$ creep less than polypropylene and polyethylene which are used above their $T_g$.

In drawn materials such as fibres, tapes or the ribs of drawn geogrids crystalline and amorphous regions become more oriented. This makes them stronger and more resistant to chemicals and weathering. The 'tie' molecules are retained and govern the mechanical properties such as creep.

The size or diameter of the basic element of a geosynthetic also influences the durability. A thick fibre or rib is less susceptible to superficial oxidation or weathering than a fine filament, not only because of the lower surface area but also because oxygen takes time to diffuse inwards and additives cannot easily migrate outwards. These factors have to be considered in evaluating the results of different durability tests.

The principal polymers used in geosynthetics are described below. Further details are given in ISO TR 13434.

- **Polypropylene** (PP) is a thermoplastic semi crystalline polymer used principally in fibrous form in geotextiles. Its structure consists of methyl groups attached to a carbon chain polymer backbone. The tertiary carbon is sensitive to oxidation so that stabilizers are added to prevent oxidation during manufacture as well as to improve long-term durability and UV stability. Commercial PP typically contains a small percentage of polyethylene.

- **Flexible polypropylene** (fPP) is an amorphous co-polymer of propylene and ethylene used for geomembranes. It is not a blend, but a reactor product using a proprietary catalyst. Its characteristics can be modified over a wide range by changing its molecular structure. It is flexible and its broad melting transition allows it to be thermally welded with a wide range of welding equipment. It is more sensitive than PP to oxidation and UV photo oxidation and therefore requires correct stabilization.

- **Polyethylene** (PE) is a semi crystalline thermoplastic. The high and medium density forms (HDPE, MDPE) are used in geobarriers and geogrids and have good chemical resistance to not oxidizing media. The linear low-density form (LLDPE) used in geomembranes and as a coating for geogrids is pliable, easy to process and has good physical properties, but is less chemically resistant. Stabilization is needed against weathering and oxidation. Durability is strongly dependent on kind of polymer, morphology, residual heavy metals like Cr and Ti from catalysts and chemical impurities from polymer processing like vinyl- and keto-compounds. Some grades of HDPE are especially susceptible to environmental stress cracking.

- **Polyesters** are a group of polymers, of which polyethylene terephthalate (PET) is commonly used as a fibre/ yarn in geotextiles and as fibre/yarn and strips for geogrids. Being below its $T_g$ of about 50 °C for the slow durability processes, PET offers good mechanical properties and chemical resistance to not alkaline media. The fibres are used in large volumes as reinforcements in car tyres, so that for reasons of price the same fibres are used in geosynthetics with no change in specification. The ester links in the polymer chain hydrolyse slowly in presence of liquid water or water vapour, leading ultimately to rupture of the chain and a corresponding reduction of molecular weight. Since this can occur in the bulk or the surface of the fibre alike it is called "internal hydrolysis" in contrast to "external hydrolysis" which is a different chemical process by etching of the fibre surface ester bonds by highly alkaline media (e.g. hydroxyl ions and ammonia). Polyethylene naphthalate (PEN) polyester products are less prone to hydrolysis then PET but more sensitive to photo oxidation by weathering.

- **Polyvinyl chloride** (PVC) is again a member of a group of vinyl-based resins used in geomembranes and as a coating for geogrids. Fundamentally a rigid polymer, by blending it with plasticizers (PVC-P) and fillers it can be made more or less flexible. Some organic liquids have the same effect. If the plasticizer migrates , e.g. due to heat, extraction by liquids or evaporation or degrades e.g. by heat or ultraviolet light or hydrolysis, PVC becomes brittle and dark in colour. The durability of PVC varies widely according to the formulation, stabilizer content and type (kind and molecular mass) of plasticizer used.

- **Polyamides** (PA) are thermoplastics used as fibres or wires in geotextiles. There are different types, numbered PA6, PA6.6 etc according to the kind of monomers used. PA fibres have good mechanical properties, resistance to wear and abrasion and offer a combination of properties including ductility, wear and abrasion resistance and low frictional properties. PA absorbs water, which makes it more flexible, reducing the $T_g$ of 40-60 °C to temperatures approaching service temperatures. It has limited resistance to acids, oxidation, hydrolysis and weathering which has to be addressed by means of suitable stabilizers.

- **Aramid** is a synthetic fibre produced from aromatic polyamides and used in geosynthetics. It has very high stiffness and strength at comparatively low density and has low tensile creep. It is sensitive to photo oxidation by UV, absorbs moisture and can be sensitive to hydrolysis by acids.

- **Polyvinyl alcohol** (PVA) is made by hydrolyzing polyvinyl acetate. Fibres are used in geotextiles. They have a high strength and stiffness and a good resistance to alkalis (e.g. concrete), lower concentrations of acids, and oils. The $T_g$ is in the range of 80 °C and thus needs to be considered for durability by Arrhenius extrapolations, as is the case e.g. for PET and PA.

- **Ethylene propylene diene monomer (EPDM)** is an elastomer used in geomembranes. It consists of saturated polymeric chains made up of ethylene and propylene molecules copolymerized with a diene monomer, e.g. ethylidene norbornene (ENB) which provides cure sites for vulcanization i.e. cross linking by sulphur or peroxides. Generally its resistance to ozone is good, depending on the residual amount of double bonds of the diene component and thus on curing. The properties of EPDM are strongly influenced by the ratio of the content of propylene, ethylene and diene monomer. Welding and repair of welds is to be done by specialists. Stabilizers are added to enhance the resistance to oxidation and UV.

- **Ethylene interpolymer** alloy is another elastomer used in geomembranes and is stabilized in the same manner as EPDM.

- **Chlorinated polyethylene** is used in geomembranes and more widely in hot water pipes. Its lack of crystallinity compared with PE makes it relatively flexible. Stabilizers are added to further enhance the resistance to oxidation, being already higher than for HDPE and to photo oxidation by UV.

- **Chlorosulfonated polyethylene (CSPE)** is an elastomer used in geomembranes. It has minimum crystallinity, is flexible, and is relatively resistant to oxidation and UV. Welding is problematic due to the cross-linking like for other elastomers like EPDM.

- **Bitumen** comprises modified bitumen (MB) and oxidized bitumen and is used to provide water tightness in geomembranes. Modified bitumen includes up to 25% synthetic elastomer which increase elasticity, fatigue resistance and ageing. Sensitivity to UV and oxidation can be adjusted by suitable additives.

- **Polystyrene** is mainly used in extruded form (XPS) or as an expanded geofoam (EPS). Suitable stabilizers are required to provide sufficient resistance to oxidation and photo oxidation.

- **Glass** fibres are used in geogrids for asphalt reinforcement. Since they are not polymers and since their required lifetimes are relatively short their durability is not considered in this publication. Degradation would be expected to result from crushing of the fibres during installation and trafficking, leading to local fracture of the fibres, progressive loss of stiffness and weakening of the geogrid. Chemical degradation results from the sensitivity to hydrolysis, which is strongly depending on the grade of the glass. Lifetime could be predicted by fatigue testing under simulated conditions. Extraction of geogrids by dissolving the asphalt may be possible, followed by measurement of their retained strength. Long-term rupture of glass fibres is known to occur under the combined presence of mechanical stress and acids,Other materials are continuously being intro-duced, for example elastomeric membranes and ABS in drainage materials. This book is limited to the principal polymers in use at the time of writing, though the general methodology should be applicable to any polymeric material.

Most of these materials can be recycled, and this is increasingly encouraged. Materials can be recycled immediately after processing (known as regrind), imported from other industrial processes (post-industrial resin or PIR) or collected from consumers (post-consumer resin or PCR). The use of small percentages of regrind is commonplace (less than 10% for geomem-branes) but other forms of recycled resin are discouraged for the more demanding applica-tions of geosynthetics such as geomembranes and reinforcements, as the composition cannot be assured. Nevertheless, if geosynthetics made with fibres derived from discarded polyester drinks bottles are not already on the market then it is likely that they will appear soon.

## 2.1.3   *Additives*

The long life expectation of polymers is due above all to the development of stabilizing additives which arrest or retard degradation and extend polymer life by ten or a hundred times. An additive is any substance which is added in a small quantity to a polymer to modify its properties for a particular purpose, such as:
- Controlling polymerization: chain growth regulators, accelerators, compatibilizers, cross-linking promoters
- Improvement of processing: flow promoters, plasticizers, slip agents, lubricants, thixo-tropic agents, release agents
- Improving resistance to degradation both during processing and service: thermal stabi-lizers, UV stabilizers, metal deactivators, acid scavengers
- Improvement of mechanical properties: nucleating agents, compatibilizers, impact modifiers, cross-linking agents, plasticizers, fibrous reinforcements
- Improvement of product performance: anti-statics, blowing agents, friction agents, flame retardants, plasticizers, smoke suppressants
- Improvement of surface properties: adhesion promoters, anti-fogging agents, anti-block-ing agents, surfactants, anti-wear additives
- Improvement of optical properties: nucleating agents, pigments and colorants, bright-eners

• Reduction in cost: extenders, particulate fillers.

Note that some additives such as plasticizers appear under more than one heading and therefore affect more than one property. By no means all additives listed here apply to geo-synthetics: the list is deliberately to illustrate the large number of different additive types in the polymer industry. The additive content of any one polymer used in a geosynthetic is generally a commercial secret. Take it that the user will not know what his product contains beyond its general polymer type, colour and shape.

Additives have to be stable at the high temperatures, pressures and shear rates experienced during processing. They must be sufficiently resistant to impurities likely to occur in the product. They must be compatible with the other additives used. They must not evaporate or sublime at high temperatures or migrate and leach out when the product is immersed in water or other common liquids. The price must be acceptable for the application in hand.

A useful list of additives is given in ISO/TS 13434. This is a summary of the classifications given there:
• Antioxidants resist degradation by oxidation. This can be achieved in a variety of ways and is more closely described in chapter 3.1 on polyolefins.
• Acid scavengers protect the polymer from acids resulting from catalyser residues or e.g. reaction products of antioxidants.
• Metal ion deactivators form inert complexes with transition metal ions that would otherwise accelerate oxidation.
• UV stabilizers provide resistance to ultraviolet light by introducing chemical compounds or carbon black that absorb the light, or by suppressing the resulting photo-oxidation.
• Plasticizers make a rigid polymer flexible, such as PVC.
• Lubricants assist in the production process such as in calender roll release of a sheet material.
• Mineral fillers such as clay and calcium carbonate are added to reduce cost and can improve the mechanical properties of the polymer.
• Scrims are usually open-weave polyester fabrics inserted between the individual plies of a geomembrane to provide reinforcement.

Table 2.1 shows a generalized table of commonly used geosynthetic polymers and the additives used Koerner, 1993.

Table 2.1 Percentage (in weight) composition of commonly used geosynthetic polymers.

| Polymer type | Basic polymer | Various additives | Carbon black or pigment | Filler | Plasticizer |
|---|---|---|---|---|---|
| Polyethylene | 97 | 0.5-1.0 | 2-3 | 0 | 0 |
| Polypropylene | 96 | 1-2 | 2-3 | 0 | 0 |
| PVC (unplasticized) | 80 | 2-3 | 5-10 | 10 | 0 |
| PVC (plasticized) | 35 | 2-3 | 5-10 | 25 | 30 |
| Polyester | 97 | 0.5-1.0 | 2-3 | 0 | 0 |
| Polystyrene | 97 | 0.5-1.0 | 2-3 | 0 | 0 |
| Chlorosulphonated polyethylene | 45 | 5-7 | 20-25 | 20-25 | 0 |

**Summary**
- The essential features of a geosynthetic required for life prediction are as follows. They should be restated when the predicted life is presented in any assessment document: Identification of the product
- type of polymer (composition, morphology, additives, coatings)
- physical structure of the geosynthetic: e.g. thick or thin fibres forming a woven or nonwoven fabric, extruded geogrid, coated fibrous strip, geosynthetic clay liner, continuous sheet.
- joints forming part of the structure of the geosynthetic (e.g. woven, welded, integral)
- mechanical and hydraulic properties (depending on function)
- thickness or mass/unit area

# References

*Chapter 2.1*
- Heibaum, M; Fourie, A., Girard, H., Karunaratne, G.P. et al (2006): Hydraulic Applications of Geosynthetics. Proceedings of the 8th International Conference on Geosynthetics (8ICG), Yokohama, Japan, September 18-22, 2006.
- Kim, Y. C. editor (2009): Handbook of coastal and Ocean Engineering. World Scientific Publishing.
- Koerner, R.M., Daniel, D.E. EPA-QA-QC for Waste Containment Facilities CH3.pdf; EPA/600/SR-93/182.
- Koerner, R.M. (2005): Designing with geosynthetics. Fifth edition. Pearson Prentic Hall, New Jersey, USA.
- PIANC MarCom 56, The Application of Geosynthetics in Waterfront Areas, Report of the working Group 56 of the Maritime Commission, Brussels, 2011.
- Pilarczyk, K. (2000): Geosynthetics and Geosystems in Hydraulic and Coastal Engineering. Balkema, Rotterdam, 2000.

- Saathoff, F. (2001): Geosynthetics in geotechnical and hydraulic engineering. Geotechnical Engineering Handbook, Volume 2: Procedures. Ernst & Sohn Verlag Berlin 2003, page 507-597.
- Van Santvoort, G. (1994): Geotextiles and geomembranes in civil engineering, Taylor & Francis.
- Van den Heuvel, C.J.M. c.s., Molecular changes of PET yarns during stretching measured with Rheo-Optical infrared spectroscopy and other techniques, Journal of Applied Polymer Science, 1993.

## 2.2    Defining the environment

J.H. Greenwood, W. Voskamp

### 2.2.1    Applications

The applications of geosynthetics are many and various. As pointed out in chapter 1, not all are long-term, nor do they require long-term durability or life prediction. The International geosynthetics Society lists as main application areas:
- Unpaved Roads
- Road Engineering
- Railroads
- Walls
- Slopes
- Embankments on Soft Soils
- Landfills
- Wastewater Treatment
- Hydraulic Engineering Projects
- Drainage and Filtration
- Erosion Control
- Agricultural Applications.

The European Standards identify as application areas:
- EN 14249: roads and other trafficked areas
- EN 13250: railways
- EN 13251: earthworks, foundations and retaining structures
- EN 13252: drainage systems
- EN 13253: external erosion control systems
- EN 13254: reservoirs and dams
- EN 13255: canals
- EN 13256: tunnels and underground structures
- EN 13257: solid waste disposals
- EN 13265: liquid waste containments

Some examples of these applications with their *principle durability requirements* are:

**Roads: separation layer under sub base**. Nonwoven or woven geotextile. *Principal durability requirement: resistance to trafficking damage. Installation damage needs also to be considered.*

Fig. 2.13 Road construction including positioning of geotextile separator (courtesy Fibertex Nonwovens A/S).

Fig. 2.14 Road construction including positioning of geotextile separator (courtesy BECO Bermüller & Co. GmbH / Bonar Geosynthetics).

**Railways: separation layer under ballast.** Nonwoven or woven geotextile. *Principal durability requirement*: resistance to trafficking damage and abrasion. Installation damage needs also to be considered.

Fig. 2.15 Railway construction: laying of track over ballast and separator; geotextile as separator (courtesy Fibertex Nonwovens A/S).

Fig. 2.16 Railway construction: laying of track over ballast and separator; geocomposite as separator and reinforcement (courtesy Huesker Synthetic GmbH).

**Surface erosion control of slopes**. Geomat or geogrid, to limit soil loss and enhance vegetation growth. *Principal durability requirement: resistance to weathering.*

Fig. 2.17 Surface erosion protection geomat (courtesy Colbond bv.).

**Vertical drains.** Prefabricated drains placed vertically in the soil, for example at the edge of a motorway embankment, allow a saturated subgrade to consolidate faster, provide subsequent drainage, and additional mechanical support. *Principal durability requirement: resistance to compressive creep, general chemical resistance. Installation damage needs also to be considered.*

Fig. 2.18 Placing vertical drains (courtesy Cofra bv.).

**Retaining walls.** Extruded Geogrids, coated geogrids, welded geogrids, high strength woven geotextiles or high strength geocomposites enable a structure to be built occupying less space and in less time. The structure requires less or no additional backfill and thus incurs less transport costs, and is stable against earthquakes. Walls can be built almost vertically and faced with concrete or natural stone or vegetation. *Principal durability requirement: resistance to damage to the connection reinforcement/cover, creep, general chemical attack and in some cases long-term weathering. Installation damage needs also to be considered.*

Fig. 2.19 Geogrid wall formwork (courtesy Tensar International bv.).

Fig. 2.20 Geogrid wall with concrete block facing (courtesy Tensar International bv.).

Fig. 2.21 Reinforced slope (courtesy Huesker Synthetic GmbH.).

Fig. 2.22 Reinforced slope (courtesy Huesker Synthetic GmbH.).

**Road and railway construction over soft ground or old mine workings**. High-strength geogrids or high-strength geotextiles. *Principal durability requirement*: resistance to creep, trafficking, general chemical attack and weathering during construction. Installation damage needs also to be considered.

**Tunnel lining**. Geomembranes. *Principal durability requirement*: general chemical durability including leaching and the quality of welded seams. Fire resistance is also critical, though this does not form part of life prediction.

Fig. 2.23 Lining of a tunnel, geomembrane (courtesy GSE Lining Technology, LLC).

Protection of concrete dams. Geomembranes. *Principal durability requirement:* resistance to long-term weathering, general chemical durability including welded seams.

Fig. 2.24 Lining of a reservoir (courtesy GSE Lining Technology GmbH).

**Lining of reservoirs and canals.** Geomembranes. *Principal durability requirement*: resistance to long-term weathering, general chemical durability including resistance to leaching and the quality of welded seams.

Fig. 2.25 Lining with a HDPE geomembrane (courtesy NAUE GmbH & Co. KG).

**Marine structures.** Nonwoven geotextiles formed into tubes or bags and filled with sand to limit coastal erosion caused by coastal currents, or to protect the beach. *Principal durability requirement: resistance to long-term weathering, creep, abrasion and water.*

Fig. 2.26 Geocontainer (courtesy NAUE GmbH & Co. KG).

Fig. 2.27 Geotextile bags as beach protection (courtesy NAUE GmbH & Co. KG).

Fig. 2.28 Fascine mattress as filter layer

Fig. 2.29 Filter layer on dike.

**Landfills.** Geotextiles are used as liners, drainage mats and protection layers. *Principal durability requirement*: resistance to oxidation and hydrolysis, weathering, creep. Installation damage needs also to be considered.

Fig. 2.30 Landfill basal liner system (courtesy NAUE GmbH & Co. KG).

Fig. 2.31 Landfill cover system (courtesy GSE Lining Technology GmbH)

## 2.2.2    The environment below ground (excluding landfills)

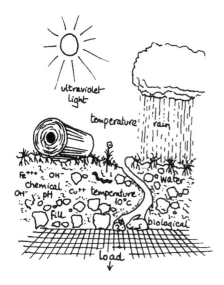

Fig. 2.32 Environment below ground (courtesy J. Greenwood).

Most geosynthetics are destined to be buried in the soil and covered by a topsoil or backfill. Topsoil, extending to a depth of up to one metre, is a mix of particles of weathered rock and humus produced by decaying organic material, with 20-60% of voids which are generally filled with air. It is biologically active. The underlying sediments consist of particles of varying size and angularity formed by the physical and chemical weathering of rocks. In backfills, which consist of sediments that have been excavated, mechanical damage may have crushed the larger particles, changing the particle size distribution. These newly crushed particles are more angular, having not yet experienced the weathering and rounding that comes with age, and their edges may be sharp enough to cut a geotextile or puncture a geomembrane. Notches can initiate crack growth in a geomembrane, particularly environmental stress cracking, and heavily indented material will form a preferential site for chemical attack.

The particle size is determined by sieving. Silt, sand and gravel are defined as containing particles of between 0.002 and 60 mm in diameter. These are generally rounded by physical weathering and transport abrasion in water and ice. Clay is defined as having particles less than 0.002 mm in size, while some clays consist of long narrow particles which cannot be characterized by one dimension alone.

The soil may be permanently or intermittently saturated with water, unsaturated but with a high humidity, or dry. In wet climates the soil drains downwards, drawing soluble materials to lower levels, while in dry climates moisture evaporates at the surface, the soluble materials being drawn upwards and deposited. In finely grained soils such as clays permeation is very slow and the water effectively immobile. The water content of an unsaturated soil can be determined by drying and weighing, but for the purpose of assessing its chemical reactivity it is described by the local relative humidity.

This water may contain dissolved chemicals, making it acid or alkaline. The measure of this is the pH, a scale which is neutral at 25 °C at pH 7 and where acids have lower numbers and alkalis larger ones. The pH of many top soils lies in the region 5.5 to 7, i.e. slightly acid, while that of the underlying sediments in Europe generally lies between 4 and 9. The most acid natural soils are found in peat moorland (pH levels between 3.5 and 4.9) and in areas affected by acid rain, the most alkaline in regions of limestone.

These dissolved chemicals may include transition metal ions such as iron or manganese, which can act as catalysts for oxidation. Bentonite, and cement used in civil engineering constructions, can lead to local alkaline areas with pH values of 8.5 to 10, while if the soil is treated with lime the pH can be as high as 11 or treatment with cement leads to pH of 12.5 or even higher. Such high levels generally decrease with time as the excessive alkalinity is carbonated by connection of $Ca(OH)_2$ with $CO_2$ to $CACO_3$. Measurements behind block concrete walls increase to a peak of 8.5 after two years and then decrease (Koerner et al. 2002), showing that while extreme conditions may prove significant, they may not represent the long term environment.

Contaminated soils contain a far wider range of pH and metal ions, depending on the industrial history of the site, and the same is true of mining waste. Some wastes are radioactive, but life prediction in such cases requires special information and is not covered in this book.

In the disturbed soil that will surround any newly installed geosynthetic the concentration of oxygen is the same as that in air, 21%. Where air circulation is then prevented and where oxygen has been absorbed by biological or chemical processes the concentration will diminish, even reaching zero. The range 5 to 15 % is given by Hsuan et al (2008) and a level of 8% was chosen for tests by Elias et al (1999). The $CO_2$ content in soil is markedly higher than in air (0.04%) and is often > 0.2 %. There may be significant concentrations of other gases such as methane, ammonia or hydrogen. Soil air generally has a relative humidity close to 100%.

It is assumed that there will always exist a thin layer of adsorbed fluid (water) on the solid particles in which the chemical reactions will take place. Within this layer the concentration of active molecules or ions is higher than in the bulk fluid. It may also occur that a transition ion acts as a catalyst for the reaction, such as ferrous or ferric ions in the oxidation of polyolefins.

The temperature in the soil surrounding the geosynthetic will vary with depth. At the surface it will vary with the cycles of day and night, winter and summer, while at depths of several metres it will approach a uniform mean. In Northern Europe this mean is close to 10 °C at depths of 10 m or more. In deep tunnels such as the St Gotthard base tunnel in Switzerland, temperatures of 40°C have been recorded. Similar soil temperatures have been observed in daytime behind the face of block walls, and since high temperatures can accelerate chemical and physical processes disproportionately, particular attention should be paid to these excessively hot periods, which have to be weighted with the appropriate Arrhenius factor for assessing the contribution to lifetime. A low soil temperature leads to a lower rate of reaction, but can also cause segregation of components (blooming, precipitation of crystalline particles) resulting in lower effective stabilizer concentrations. Very cold temperatures can make polymers stronger but more brittle, such as has been observed for a needle punched nonwoven polypropylene geosynthetic in frozen wet soil (Henry and Durell 2007). A greater problem with the design of geosynthetic structures at sub-zero temperatures is that frozen soil can act quite differently to unfrozen. This is a more general question that is separate from the durability of the geosynthetic alone and is not dealt with here.

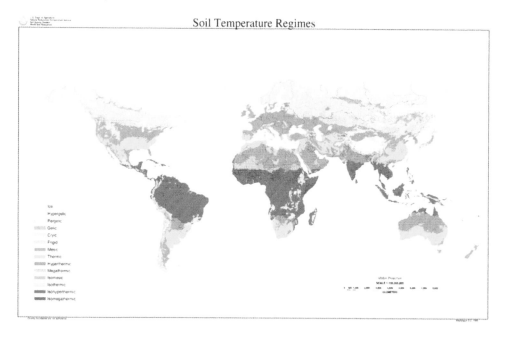

Fig. 2.33 Soil Temperatures Regimes (source: www.nrcs.usda.gov)

Figure 2.33 shows the soil temperature classes used worldwide. The following are those found in central Europe:

**Frigid**

The mean annual temperature is lower than 8°C and the difference between mean summer (June, July and August) and mean winter (December, January and February) soil temperature is more than 6 °C.

**Mesic**

The mean annual soil temperature is between 8 °C and 15 °C, and the difference between mean summer and mean winter soil temperatures is more than 6 °C.

**Thermic**

The mean annual soil temperature is between 15 °C and 22 °C, and the difference between mean summer and mean winter soil temperatures is more than 6 °C.

These temperatures are measured at a depth of 50 cm below the surface (or, in soil terms, at a densic, lithic or paralithic contact if this is shallower). Most geosynthetics are located at a greater depth.

The durability of the geosynthetic product for a specific project needs to be based on the actual temperature regime at the location of the project. In central Europe, for example, the mean annual temperature is between 8 °C and 15 °C and a reference temperature of 10 or 15 °C is commonly used.

## 2.2.3   Above ground

While most geosynthetics will spend their life underground, during installation they may be temporarily exposed on the surface. There are also a number of applications, such as at the tops of slopes of reservoirs, in erosion control and in reinforced embankments, in which at least part of the geosynthetic is permanently exposed to weathering.

Ultraviolet light, is well known as the principal cause of degradation in polymers, and this is explained more fully in chapter 3.3. Ultraviolet light, defined as having wavelengths of between 200 and 400 nm, forms 5-10% of sunlight, cloud or no cloud. This proportion increases at high altitudes and close to the poles. The photons of ultraviolet light, which have higher energies than the rest of the visible spectrum, are capable of breaking the molecular bonds within the polymer. The broken bonds form radicals which recombine with oxygen in the air to form reactive peroxiradicals which initiate more complex reactions (photo-oxidative reactions). At the same time the whole spectrum of sunlight raises the temperature of the polymer, and this together with other factors such as rain, pollutant oxides of nitrogen and sulphur, and ozone together may degrade the geosynthetic in a very complicated manner.

For the purposes of life prediction we need to know the annual exposure to ultraviolet light, known as the radiant exposure, which is generally available from meteorological data for the planned location. Figure 2.41 shows a world map of the radiant exposure. If the radiant exposure is not known it can be estimated as 5-10% of the global solar radiation.

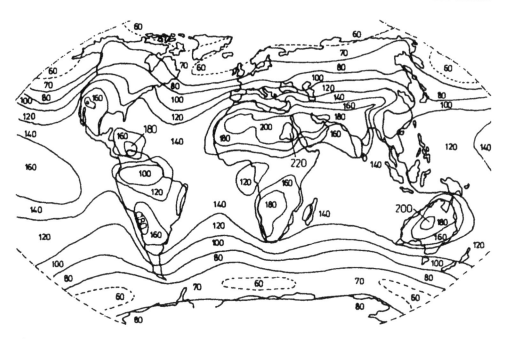

Fig. 2.34 Map of global radiant exposure in kcal/cm² or kly, Veldhuijzen van Zanten, 1986. 100 kly is equivalent to 4.2 GJ/m² /year.

Meteorological data should also be available for the local air temperature and the rainfall. Sunlight can however raise the temperature of an exposed black surface such as a geomembrane many degrees higher than the air temperature, as will be familiar to anyone leaving a car for a few hours in hot sun. Rain will cool exposed surfaces but will also help to leach out soluble material that has migrated to the surface such as additives or the products of reaction.

Animals – with the exclusion of humans – have so far not proved a significant hazard to geosynthetics, although birds have been known to sample specimens of geotextiles deliberately exposed for testing. Evidently geotextile fibres make good nests.

*Checklist for defining the environment with regards to the durability assessment of geosynthetics*

**Below ground**
- particle size distribution and angularity , type of soil/main kinds of minerals
- organic content
- acidity/alkalinity (pH)/ salt content
- amount and kind of metal ions present
- partial pressure of oxygen and other aggressive gases
- moisture content , information on soil water
- yearly temperature course

**Above ground**
- intensity of ultraviolet radiation (yearly course)
- yearly temperature course
- amount of rainfall
- type of climate

## 2.2.4   *Landfills*

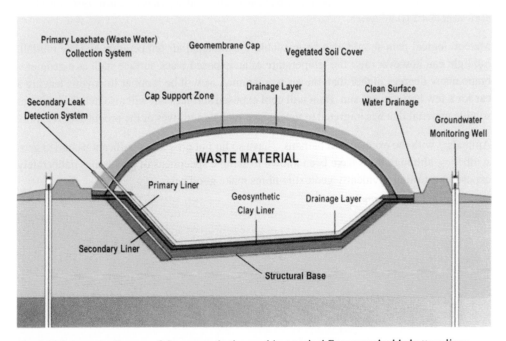

Fig. 2.35 Schematic diagram of the geosynthetics used in a typical European double bottom liner landfill.

Landfills are built for the containment of municipal solid waste as well as for industrial fluids such as mine tailings. They make use of all kinds of geosynthetic, including geomembranes, geosynthetic clay liners, woven and nonwoven geotextiles and drainage geocomposites. The principal component is an impermeable barrier made by welding together sheets of polymeric geobarriers (GBR-P). Above this there will be a layer of geotextile to protect the BNR-P from damage both during construction and from sharp stones in the mineral drainage layer. In addition there may be a filter layer to allow liquid emanating from the waste ("leachate") to be collected and treated remotely. The landfill will be covered with a further impermeable geobarrier, possibly also with a drainage layer to conduct away water from the soil above. The cap should allow for the collection of methane, which rises to the top as it is lighter than air. A drainage geocomposite for the collection of gas is therefore recommended.

The composition of municipal solid waste (MSW) is not only very variable by its nature, but will change with time in the landfill. The basic constituents of the leachate comprise volatile fatty acids and other organics, inorganic salts with traces of heavy metals, and surfactants (Rowe, Islam and Hsuan 2008). Monitoring of municipal solid waste streams in the UK also indicates that they are dominated by the biological processes responsible for the degradation of organic compounds. Legislation is in hand to reduce the content of biodegradable waste, and this should lead to a reduction in landfill temperature and thereby the rate of oxidation. On the other hand the traces of transition metal ions, notably copper, manganese and iron, can act as catalysts for oxidation. Needham et al (2004) state that landfills managed using typical current approaches may take hundreds, if not thousands, of years to stabilize.

According to Needham et al (2004), quoting Barone (1997) and Rowe (1998), the maximum temperature, observed for leachate heads of 6 m or more, is 60 °C. Other observations on landfills operating as bioreactors in the USA and UK suggest a maximum temperature of 30-35 °C which should eventually decrease, although there is insufficient information to predict the timescale at which this is likely to happen. The long-term average temperature of dry landfills or those with a low biodegradable content should be in the range 15-20 °C.

The same authors state that there will be an absence of oxygen during the methanogenic phase, although this will not change the depletion of antioxidants by processes such as leaching. The availability of oxygen will also be limited on the underside of the geomembrane liner if it is in close contact with, for example, clay. In general full oxygen availability is assumed (i.e. 21% as in air) such that the resulting predictions can be regarded as conservative.

The *principal durability requirements* for geosynthetics used in landfills are:

**Geomembranes (GBR-P)**, including welded seams: chemical resistance including resistance to leaching, resistance to environmental stress cracking, resistance to oxidation, resistance to weathering during construction.

**Geotextiles:** resistance to installation damage, resistance to oxidation, chemical resistance.

**Drainage geocomposites:** resistance to compressive creep and long-term shear, resistance to oxidation, resistance to installation damage, chemical resistance.

**Geosynthetic clay liners:** resistance to desiccation, resistance to freeze-thaw, resistance to ion exchange, chemical resistance, resistance to long-term shear, resistance to oxidation, resistance to leaching.

## References

*Chapter 2.2*

- Veldhuijzen van Zanten, R., Geotextiles and geomembranes in civil engineering, page 231, Balkema, 1986.
- Koerner, Hsuan and Koerner 2002, Koerner, G.R., Hsuan, Y.G., and Koerner, R.M. "Photodegradation of Geotextiles," Journal of Geotechnical and Geoenvironmental, ASCE, Vol. 124, No.12, pp. 1159-1166, 1998.
- Hsuan Y G, Schroeder H F, Rowe K, Müller W, Greenwood J H, Cazzuffi D, Koerner R M. Long-term performance and lifetime prediction of geosynthetics, 2008. EuroGeo4, Edinburgh, UK, pp 327-366.
- Elias, V. Et al, Testing protocols for oxidation and hydrolysis of geosynthetics, FHA, Washington, Report FHWA-RD-97-144 pp 200, 1999.
- Henry K S, Durell G R, Cold temperature testing of geotextiles: new, and containing soil fines and moisture. Geosynthetics International, Vol. 14, Issue 5, 2007, pp 320-329.
- http://www.geosyntheticsworld.com/search/label/landfill.
- Rowe, R. K., Islam, M.Z. and Hsuan, Y.G. (2008) "Leachate chemical composition effects on OIT depletion in HDPE geomembranes", Geosynthetics International, 15, No. 2, 136-151.
- Needham et al, Prediction of the long term generation of defects in HDPE liners, EuroGeo 3 2004.
- Key to Soil Temperature Classes.
- (http://www.pedosphere.com/resources/sg_usa/subgroups.cfm?Order_ID=L&SubOrder_ID=E&GreatGroup_ID=D) Pop up Glossary.

## 2.3    General durability issues

J.H. Greenwood

### 2.3.1    Introduction

In this chapter we shall examine the fundamentals of life prediction. As mentioned in Chapter 1, life prediction can be based on experience, on the extrapolation of measured properties, or on tests intended to simulate the whole life of a geosynthetic by accelerated testing. Experience, while used widely in industry for products with lesser lifetimes, is simply not available for the long design lifetimes expected of geosynthetics.

Measurements have to be made for at least ten to twenty years in order to detect with sufficient accuracy the changes that will lead to failure after more than a century. Many processes occur non-linearly, showing little evidence of change in their early stages and then reaching a threshold where degradation sets in emphatically. In others one process of degradation follows another. Life prediction is not as simple as measuring the progressive wear of a car tyre to determine when it will need replacement.

## 2.3.2 *Function and end of life*

What determines the point at which a geosynthetic is said to fail? While catastrophic rupture may come to mind, in many cases the geosynthetic will change gradually, very gradually, by extension, compression or by the growth of small holes. "End of life" is a more appropriate description than failure.

End of life depends on the function for which the geosynthetic has been chosen. Geomembranes (geobarriers) generally act as impermeable barriers to water, leachates, industrial effluents or gases. The functions of geotextiles include drainage, filtration, protection, reinforcement, separation, surface erosion control and asphalt reinforcement (see ISO 10318). Some materials may have more than one function, such as the nonwovens knitted with parallel yarns to provide reinforcement as well as the capacity for filtration, separation and even drainage.

The function determines the end of life. A filter or separator cannot sustain a substantial tear or rupture. A reinforcement must not rupture, but it may also cease to function correctly if it stretches by more than a certain amount, leading to sagging of a reinforced wall or a roadway embankment. A reinforcement intended to resist seismic loads will reach its end of life once its strength has sunk below a required level, even though it may never have experienced those loads. A drainage composite must retain a certain capacity for flow, which is represented by a minimum thickness of the core.

To make a numerical life prediction the end of life has to be defined as a number. Vague general statements are not sufficient. Examples of a quantitative and measurable criterion are as follows:
- 50% reduction in strength
- 2% creep strain of a reinforcement measured from the date of completion of construction work
- 25% reduction in core thickness for a drainage composite
- 10% increase in permeability of a PVC geomembrane
- visible holes in a geomembrane.

In the case of a polymeric geomembrane (Needham et al 2004) the end of life criterion may require further definition in terms of the number and size of holes, according to:

- whether the geomembrane is a single or double liner or a composite, in which case the quality of contact between the geomembrane and the underlying mineral liner has to be considered
- the leachate source is persistent, leachate head, and the chemical nature of the leachate
- the sensitivity of the site to groundwater quality.

Thus the end of life criterion can vary from one site to the next.

The statement of lifetime, or of long-term durability, will refer directly to the end of life criterion used.

### 2.3.3 *Required and available properties*

A simple statement of end of life may not reflect the complexity of some applications. ISO TS 13434 introduces the concept of required and available properties. It is clear that in the long term the geosynthetic must retain the end of life property defined above. During construction of the soil structure, however, there may be special temporary requirements. The material may need to be even stronger to resist the weight of the construction equipment, or it may require a particular resistance to weathering. If this 'required' property is plotted against time it will therefore no longer be a single value but will rise and fall according to the history of the structure in which it is used.

The 'available' property of the geosynthetic will also vary with time. For example, it may fall abruptly due to installation damage and then more slowly in the course of slow chemical attack.

The available" property must at any time be greater than the 'required' property, probably by a minimum margin. A schematic graph showing how the available and required properties may change with time is shown in figure 2.36.

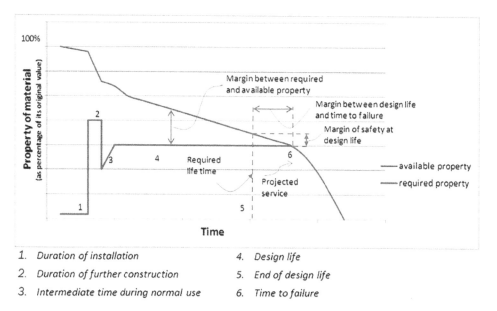

1. Duration of installation
2. Duration of further construction
3. Intermediate time during normal use
4. Design life
5. End of design life
6. Time to failure

Fig. 2.36 Available and required properties of a geosynthetic during storage and transportation, construction, backfilling and use. Duration prior to installation (storage and transportation).

Comparison between the required and available properties at every stage enables the design engineer to allow for a minimum margin of safety. This will take into account the consequences of failure and is part of soil structure design. With regard to life prediction of the geosynthetic alone this procedure is seldom followed; most such predictions are based on a single end of life criterion that does not change with time.

## 2.3.4   Design lifetimes

The purpose of life prediction is generally to provide assurance that the lifetime of the geosynthetic, as defined by its end of life criterion, is greater than its design life. In other words, the geosynthetic will continue to do its job. This design life will be set by a code such as that of the European Organisation for Technical Approvals (EOTA 1999) which defines out four standard lifetimes of 10, 15, 50 and 100 years for construction materials, according to the ease with which the material can be repaired or easily replaced. For geosynthetics we are normally considering lifetimes of 50, 100, or, as we shall see, even longer durations in applications where the geosynthetic cannot easily be replaced. Typical of these are reinforcements of block walls, filters and drainage elements in motorway embankments, materials for coastal erosion control, geomembranes in tunnels, canals, reservoirs and above all in landfills. In contrast to these there are applications where the function of the geosynthetic is only temporary. Some examples of contrasting design lifetimes are given in the box below.

| | |
|---|---|
| Separator as construction aid | 0.5-1 year |
| Separator (permanent) | 80-100 years |
| Filter in drainage applications, replaceable | 10-25 years |
| Filter in drainage applications, not replaceable | 80-100 years |
| Reinforcement in a dam against slip failure | 5 years |
| Reinforcement in retaining structures | 80-100 years |
| Prefabricated vertical drain | 1-3 years |
| Landfill liner, drainage and protection | 100+ years |
| Tunnel liner | 100+ years |

Some authorities have asked for infinite life. This would only be possible if all possible forms of degradation were shown either to be negligible or to approach some finite limit, and if it was believed that environment will not change during the relevant working life. Will the next 100 years bring global warming as predicted, or will it lead to an ice age in northern Europe such as occurred only 10000 years ago?

Design lives are stated in years; accelerated tests are measured in hours. It is therefore convenient to state here the number of hours in a year:
1 year = 8760 h
Similarly for a century, including leap years:
100 years = 876576 hours.
Since on a logarithmic scale 876576 is scarcely distinguishable from 1000000 or 106 hours (1000000 hours = 1Mh = 114 years), this number is often used in the evaluation of accelerated tests.

## 2.3.5 *Extrapolation*

The simplest – and ideal – method of life prediction is the extrapolation of existing measurements to the limit which is known to be unacceptable. The diminishing tread of a car tyre can be measured, plotted against time, and a prediction made of the date at which it will reach the legal minimum, excluding any other effects such as cracking or ageing of the rubber. Such changes in geosynthetics are very slow and many years will be needed to predict with confidence a lifetime of 100 years or more. Nevertheless, this method has the essential advantage that the environment which has caused the geotextile to degrade in is the same as that for which prediction is being made. Nobody can argue about which of the several forms of degradation predominates or whether there is synergy between them – all are present in their correct proportions and sequences. The only assumption – which may be significant in a century of predicted climate change – is that the environment remains constant.

While this method is powerful in the case of installed materials, for new materials neither manufacturer nor client will be prepared to wait for 20 years to obtain a prediction. It should be seen more as a method for monitoring – and where necessary adjusting – the predictions obtained by short-term accelerated testing. Details are given in chapter 3.8.

There are however certain general principles concerning extrapolation whilch will be mentioned here. The general principle to be used in extrapolating data is to use the simplest theory that fits the facts. Known as 'Occam's razor', this principle was first expressed by William of Occam (1285-1349), an English philosopher who practised in Oxford and Munich, and applies as much to the extrapolation of scientific data in the twenty-first century as it did to the church's philosophy in the fourteenth century.

If the results fall on a straight line, then this can be extended. If not, then a curve can be fitted. A generation ago an engineer would have used a "flexicurve", a rod of soft metal in a plastic sleeve, to extend his creep curve by hand as it looked right to him. Do not disparage such methods. There is generally no fundamental law covering degradation; most curves are empirical. Today he is more likely to use a computer-generated approximation such as the power law, which also forms the basis of the Excel Trend function. This has generally proved satisfactory.

The use of polynomial functions is discouraged, since on extrapolation they can go into reverse and predict negative creep which is clearly ridiculous. Always be on the lookout for the unreasonable or the impossible. In the example shown later in figure 3.94 one could fit and extend a straight line through the points – except that it would predict a zero and then negative stabiliser content. The operator, not the computer, is the final judge of what is a reasonable prediction.

Extrapolation should ideally not exceed a factor of ten: if the measurements have lasted for two years, then extrapolation should not be for more than twenty years. This is also demonstrated in chapter 3.4. Extrapolation beyond this point, if unavoidable, should be accompanied by an appropriate safety factor.

## 2.3.6 Reduction factors for reinforcements

While for most applications the purpose of life prediction is to ensure that the lifetime of the geosynthetic is greater than its design life, in the specific case of reinforcements a loss in strength is allowed for during the life of the reinforced soil structure. Strength is the critical property of a reinforcement; without it the soil structure would collapse. This strength diminishes with time due to a variety of influences. To allow for this the initial strength of the geosynthetic - the manufacturer's proud and quality assured number – is reduced by a number of reduction factors reflecting the stress the geosynthetic is under and the environment surrounding it. Most of these factors will themselves change with time. ISO/TR 20432 defines them as follows:
- $RF_{CR}$ : reduction factor for creep-rupture
- $RF_{ID}$ : reduction factor for installation damage
- $RF_{WE}$ : reduction factor for weathering
- $RF_{CH}$ : reduction factor for the environment, including chemical and biological attack

For applications such as railways, it may be appropriate to add a further reduction factor $RF_{dyn}$: reduction factor for dynamic loading. The German system adds a reduction factor for joints and connections (see chapter 3.7).

The reduction factors are greater than 1.0 or, if no degradation is expected, equal to 1.0. In addition there is a further factor of safety $f_s$ to take into account uncertainties. All these factors are multiplied together to yield a single factor by which the initial strength is divided to give the design strength. It might be more logical to define the reduction factors as numbers less than 1.0 by which the strength should be multiplied. Current practice is purely convention and any attempt to change that convention now would only cause confusion. Chapter 3.7 describes how these factors are to be derived from the available data.

'Initial' strength itself requires better definition. A new geosynthetic will have a quality assured characteristic strength which according to most codes of practice is defined as the strength which is exceeded in 95% of measurements made under routine product testing, or alternatively two standard deviations below the mean. The US definition of the Mean Average Roll Value (MARV) states this explicitly. The European applications standards EN 13249-13257 and 13265 (2001) define a tolerance corresponding to the 95% limit and then state that the 95% confidence level corresponds to the mean minus (or, in some cases, plus) one tolerance. The data stated in a CE marking document in accordance with these standards are nominal values, guaranteed by the manufacturer. They do not have to be in direct relation to the manufacturer's statistics. For a normal distribution 2.3% of the measurements should be less than the mean minus two standard deviations and a further 2.3% above the mean plus two standard deviations. Unusually high strengths do not worry us. Thus the assurance is actually 97.7%. In practice many manufacturers define a reinforcing product by its assured nominal strength (e.g. WonderGrid400 for a 400 kN/m product) rather than releasing the statistics of their internal quality control. It is reasonable to use this as the initial strength.

### 2.3.7    *Expression of life prediction*

Life prediction will generally take the form of an assurance that, within the limits of current knowledge, a geosynthetic will continue to fulfil its function throughout its intended design life. This assurance will be part of the initial design to meet a specific specification or legislation. In Europe CE-Marking includes an assessment if the material is to be marked as durable for environments or durations that lie outside set limits, which currently include all applications with design lives of longer than 25 years. The assessment will state that material X, in environment Y, is expected to fulfil its function according to end-of-life criterion Z, for a minimum lifetime T. For general marketing purposes, or to counter competitors' claims, a more general statement may be required.

For reinforcement design the life prediction take the form of reduction factors by which to reduce the long-term strength of the reinforcement in design, together with predicted strains, again for a material X in and environment Y for a lifetime T.

## 2.3.8   Safety factors

No prediction can be exact. Uncertainties can arise from the variability in properties, the precision with which they are measured, as well as from more indefinable matters such as the choice of method for prediction. Safety factors can be based on statistically measured properties or on codes and procedures. They may be expressed as a function of the end-of-life criterion, such as a 50% margin on strength, as the difference between the required and the predicted available property, or as a margin on design life. This will be discussed in chapter 3.7.

The safety factors defined here only cover the geosynthetic itself, not the design of the soil structure. For example, collapse of a reinforced wall could lead to loss of life as well as disruption to road or rail traffic, while leakage of a pond containing toxic chemicals could lead to contamination of the water supply. Degradation may be gradual, allowing for detection and remediation, or it may be sudden. The civil engineer may choose to apply his own safety factor to allow for these eventualities as part of his design. Lifetime assessment as set out in this book should be kept separate from soil structure design and the consequences of its possible failure.

## 2.3.9   Accelerated testing

### 2.3.9.1   Introduction

The most common methods for predicting lifetime in industry are statistical and depend on the presence of a large amount of historical information. Extensive databases exist on the times to failure of items as diverse as microelectronic components and construction materials such as doors, windows and drainpipes. Common to these databases is that many similar components have been used in similar environments, that the lifetimes are not large, and that the changes to the components in the meantime have not been so extensive as to make the information obsolete. Also essential is a means of systematic recording and analysis. Human beings themselves are not exempt: the predicted life expectancy of males and females, smokers and non-smokers alike, is readily available on the internet.

This information does not yet exist for geosynthetics. Design lifetimes of a hundred years or more are very long in comparison with most applications of plastics and polymers and, more importantly, they are longer than the geosynthetics themselves have been in existence. Bakelite, the first commercial plastic, was only patented in the first decade of the twentieth century, while most of the polymers from which geosynthetics are made were invented between 1930 and 1960, and the additives which extended their durability so dramatically were developed later still. Such data as there is generally refers to earlier products and, so often, the environment in service has not been measured. In this situation the only option is to base life prediction upon accelerated testing.

Accelerated testing comprises numerous methods intended to obtain or predict failure within a time acceptable to the manufacturer, and evidently much shorter than the design life. The methods fall into three groups:
- Acceleration by increasing frequency or eliminating 'dead' time;
- Acceleration by increasing the severity of the degrading agent, such as load or chemical concentration;
- Acceleration by increasing the temperature: this can accelerate not only chemical degradation but also physical processes such as creep, migration, leaching and evaporation.

In selecting any accelerated procedure the following critical rules apply:
- The mechanism of degradation during the full range of accelerated testing must be unchanged and must be as close as possible to that expected to occur in service.
- There must be no change in the physical structure of the material (morphology) over the full range of temperature and time, for example no rubber to glass transition, no formation of barrier layers which might retard the degradation process, and no surface cracking that might accelerate it, unless it can be demonstrated that the change has no effect on the property being measured. The appearance of the samples or fracture surfaces after accelerated testing shall give no evidence of such a change.
- Degradation should be due to one dominant factor.
- There must be a measurable change in property with time.

The last two points are discussed further in the following sections.

### 2.3.9.2   Selecting the dominant factor

While the environment encompasses various different factors that could cause degradation, most life prediction is based on the assumption that one dominant factor is responsible. Accelerated ageing therefore generally increases this one factor only.

Be clear that this is a generalisation. Verdu et al (2007) have described how one should choose accelerated conditions that simulate natural ageing, and has recommended the measurement of more than one parameter to ensure that the acceleration is valid. Acceleration is generally based on the kinetics of molecular changes which may be separated by several steps from the macroscopic properties of interest. For example in hydrolysis the acceleration relates to a chemical reaction which then reduces the chain length. This then leads to loss of strength. Each step may be governed by its own mechanism, such as chemical rate kinetics or fracture mechanics, and each depends on the accelerating parameters in its own way. Transport mechanisms such as migration and diffusion may themselves be dominant and have to be considered. The more complex the mechanism, the more difficult it is to simulate, particularly in the case of oxidation.

Oxidation has been accelerated by using raising the temperature and the oxygen pressure simultaneously (chapter 3.1), but even this requires multiple tests and three-dimensional correlation. Life prediction with multiple accelerated factors, sometimes called multi-stress ageing, has so often proved intractable.

### 2.3.9.3  Modes of degradation

Degradation can occur immediately, gradually, or suddenly after a long period of time. These are termed "Modes of degradation" and numbered 1, 2 and 3 respectively. They are depicted in figure 2.37.

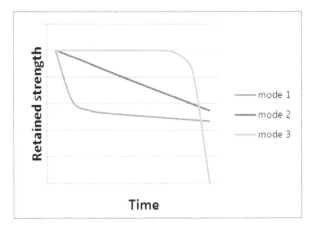

Fig. 2.37 The three modes of degradation.

Installation damage is, in the context of geosynthetic lifetime, relatively immediate. There is a rapid and irreversible reduction in strength but no further change due specifically to damage. This is Mode 1. This is not strictly a life prediction issue, since there is no change with time, but it forms part of any durability assessment.

Hydrolysis of a polyester geosynthetic is an example of Mode 2. Moisture penetrates the fibres slowly and leads to a general and gradual reduction in strength. Life prediction is a matter of understanding the rate at which these changes occur and extrapolating their effects into the future.

Mode 3 is the most difficult to predict. The oxidation of a stabilized polypropylene is an example; while the stabilizer is present the strength is unchanged, but once the stabilizer has been used up the strength of the unprotected polymer can fall rapidly. In accelerated oven ageing what was a polypropylene filament one day can be found as a pile of black powder the next – though do not fear that a geosynthetic in the soil will change so fast. Life prediction must adopt a more subtle approach, such as measuring the depletion of stabilizer with time, in order to predict well in advance the time at which the material will weaken or collapse.

### 2.3.9.4   Acceleration by increasing frequency

Take a car whose lifetime is typically fifteen to twenty years. During this lifetime it will start, warm up, drive, stop, cool down, park. Doors and windows will be opened and shut, locks activated. Its actual total driving time will be less than twelve months; for the rest it stands parked and gleaming in front of its owner's house – 'dead' time. Although some forms of degradation, such as rust and the fading of paint, will continue while the car is parked, many will depend on the time driven, on the number of times the engine warms and cools, or on how often the door is slammed shut. The last three can be simulated by eliminating the dead time: driving the car continuously round a track, repeated starting and stopping, and opening and closing the door, with only the minimum necessary interval between them. Car manufacturers do this. The total time needed to simulate the driving lifetime of the car can be less than a year.

For geosynthetics most types of degradation are due to continuous exposure, such as to load or chemical environment, with no dead time. This form of accelerated testing can be used in only a few cases. Examples are:

- Weathering: since the sun shines for an average of only twelve hours a day, the rate of exposure to weathering can in theory be doubled simply by applying artificial sunlight day and night. Even then, some of the effects of weathering are due to the alternation of light and dark, wet and dry, so that accelerated weathering tests generally include a reduced period of darkness and occasional water sprays. Acceleration by this means is therefore by less than double.
- Traffic loading: this is intermittent and has been accelerated by eliminating the dead time between loads. Railway wagons have been driven continuously backwards and forwards on a track laid out over a trial array of geosynthetics to simulate a much longer period of service.
- Tidal surges: Where the function of a geosynthetic is used to resist exceptional conditions such as tidal surges and where degradation of the geosynthetic only occurs under these extreme conditions, testing could be performed under these conditions alone over a short period.

In most cases, however, other forms of acceleration apply.

### 2.3.9.5   Acceleration by increasing severity

A chemical reaction will normally proceed faster if the chemical is more concentrated. If the reaction is oxidation then it will proceed faster when the oxygen is under pressure. The same will be true if the intensity of ultraviolet light is increased, or the load is heavier. To understand the effect of these increases on lifetime, it is necessary to have a numerical relation between the severity (i.e. concentration, intensity, load) and the rate of degradation.

If there is an established relation it should be possible to define a single test under severe conditions to simulate a longer lifetime under service conditions. Clearly it is assumed that the mechanism of degradation is unchanged.

For example, it should be possible to use chemical process theory to relate the rate of degradation to chemical concentration and therefore the measured lifetime under more severe conditions to a predicted lifetime under more normal conditions.

In predicting the resistance to weathering it is assumed that the level of degradation is proportional to the radiant exposure, or accumulated dose of ultraviolet light. This is the basis for the index test EN 12224 which applies 50 MJ/m² of ultraviolet radiation at an increased intensity (Greenwood et al 1996). The rate of degradation is assumed to depend on the intensity of the radiation and the total extent of the degradation, for example the loss of strength, on the total dose of radiation experienced. The limit on intensity is generally set by the resulting rise in temperature, because the intense ultraviolet light heats the specimen until it reaches the maximum temperature acceptable for the material.

If the relation between severity and rate of degradation is not known or is statistically uncertain, then multiple tests will be needed to measure it. Creep-rupture is such an example. In this case tests are performed at a range of different high loads on separate specimens and the time to rupture is measured. A relation is then established between the load and time to rupture (the inverse of the rate of degradation), commonly that the logarithm of the time to rupture increases linearly with decreasing load. Having established this relation, its graph can be extrapolated from the short lifetimes and high loads used in testing to predict the longer lifetimes that correspond to the lower loads applied in service, with the object of defining a design load corresponding to the service lifetime (figure 2.38).

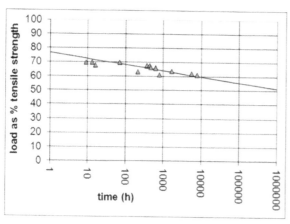

Fig. 2.38 Creep-rupture of a polyester reinforced geotextile. The measured times to failure are plotted horizontally – on a logarithmic scale – as a function of their applied loads. The relation between applied load and the logarithm of time to failure is not known, but is as- sumed to be a straight line as

drawn above. Having established this relation, it is then extrapolated to give the load that would lead to failure after 1000000 h, namely 51% of the tensile strength.

Increase in severity is sometimes combined with an increase in temperature in order to provide the degree of acceleration necessary to simulate the very long lives of geosynthetics. For example, in ISO 13438 Methods C1 and C2 oxidation is accelerated by simultaneously increasing both oxygen concentration and temperature. Creep-rupture is performed at multiple temperatures. These are described further in chapters 3.1 and 3.4 respectively. Increase in chemical concentration as well as temperature is the basis for the index tests in ISO 14030, in which specimens of geosynthetic are exposed separately to concentrated aqueous solutions of sulphuric acid and calcium hydroxide. Oxygen is bubbled into the acid to increase the likelihood of oxidation. In these cases, however, the correlation between the test conditions and long-term lifetime is less certain and the function of the index test is to eliminate any products where there is any risk of degradation under acid or alkaline conditions such as for polyamides in soils with a higher acidity than pH4 for a working life of 25 years.

### 2.3.9.6    Acceleration by increasing temperature: Arrhenius' formula

It is a fundamental experience of life, not least in cookery, that raising the temperature accelerates chemical, physical and biological processes. To make a first approximation, note that there is a widely used rule of thumb, attributed to the Dutch scientist and Nobel Prize winner Jacobus Henricus van't Hoff, by which the rate at which a process occurs doubles with each 10 °C increase in temperature. This covers the performance of many polymers over their range of use. In some cases a factor of three or four is more appropriate than a factor of two. This provides a useful tool for a first approximation.

The Swedish chemist and Nobel Prize winner Svante Arrhenius derived from thermodynamic considerations a proper theoretical relation between temperature and the rate at which a process occurs. Intended originally to cover chemical reactions, this relation has also been applied successfully to physical processes such as creep. Arrhenius' formula is used to evaluate most chemical and many physical tests accelerated by increasing the temperature.

Fig. 2.39 Nobel Prize winner Svante Arrhenius. He also derived a numerical relation between the concentration of atmospheric carbon dioxide and global warming (Elsevier,1996).

Fig. 2.40 Nobel Prize winner Jacobus Henricus van 't Hoff. He developed the rule of thumb by which the rate at which a process occurs doubles with each 10 °C increase in temperature.

Arrhenius' formula can be stated as follows:

$$A = A_0 \exp(-E/R\,\theta)$$

where A is the rate of degradation, $A_0$ a constant or pre-exponential factor, E the activation energy of the process in J/mol, R the universal gas constant (8.314 J/mol.K) and $\theta$ is the absolute temperature in K (temperature in °C +273). Note that for geosynthetics the Greek theta $\theta$ is used instead of the more common T to represent temperature, since in geotechnical engineering T is used to represent load per unit width. E relates to the fundamental molecular mechanism driving the process, such as the breaking of a chemical bond or the unravelling of a polymer chain. As $\theta$ increases, so $E/R\,\theta$ decreases, the exponential term increases and with it the rate of degradation. Values of E in the polymer field tend to lie between 30 and 110 kJ/mol; a higher value of E leads to a greater sensitivity of the rate of degradation to temperature. Such mechanisms are described as being 'thermally activated'. E is taken as independent of temperature, though it may show a weak dependence, and it can be affected by the presence of catalysts or changes in pH. Processes with high activation energies are

117

accelerated faster by an increase in temperature, something that must be borne in mind when, as with oxidation, several thermally activated processes operate simultaneously.

To obtain a straight line graph take natural logarithms of both sides of the equation to give:

$\ln A - \ln A_0 = - E / R\, \theta$.

The natural logarithm of the rate of degradation ($\ln A$ ) is then plotted against the inverse of the absolute temperature ($1/\theta$ ). Arrhenius' formula applies if the points lie on a straight line with gradient $- E/R$, from which the activation energy $E$ can be calculated.

The logarithm to base 10 ($\log A$ ) can be used instead of $\ln A$, but then the gradient is then equal to $- E / 2.3036\, R$. If time to failure ($t_f$ ) is used instead of rate of degradation Arrhenius' relation and its logarithmic form become

$t_f = t_{f0} \exp( E / R\, \theta )$

$\ln t_f - \ln t_{f0} = E / R\, \theta$.

where $t_{f0}$ is a constant in the same manner as $A_0$. Note that the minus sign before E has been removed and that the graph now has a positive instead of a negative gradient.

### 2.3.9.7   Planning Arrhenius tests

The procedure for planning a series of Arrhenius tests should include the following:
- Identify the mechanism of degradation and confirm that it will be accelerated by temperature. Identify any other potential forms of degradation and how they can be avoided or excluded by using control samples.
- Define a single property related to the end of life and the limit of its acceptability. This may be the property of direct interest, such as 50% retained tensile strength. A change in property is essential. Note, however, the following.
  - A higher retained strength (e.g. 90%) may be selected in order to shorten the test durations, provided that the time to 50% retained strength can be predicted with confidence from the results.
  - It is important that the property can be measured with sufficient precision to monitor the changes. It may be better to measure a different property such as intrinsic viscosity which can be measured with greater precision than tensile strength, provided that there is an established relation between them, yet tensile or burst strength may prove more precise than hydraulic permittivity.
- Determine the environment, such as, air, water of a standard purity or a chemical.
- Decide on the range of temperatures to be used. In general four temperatures are recommended, although three are sufficient. The maximum temperature will be that at which the mechanism of degradation or the state of the polymer changes, and the minimum temperature will be limited by the time available for testing. Temperature steps should generally be no greater than 10 °C, but may be larger for properties less sensitive to

temperature. Ideally the lowest temperature should be no more than 25 °C above the predicted service temperature, but since the design lifetime for geosynthetics is typically 100 years and the time available for testing rarely exceeds one year, this temperature interval is likely to be greater. ISO 2578 provides further guidance.

- Define the times after which testing is to take place. These times should be spaced logarithmically, for example after 100, 300, 1000, 3000 and 10000 h, since the logarithms of these times, 2, 2.5, 3, 3.5 and 4, are equally spaced. Since it is not possible at the outset to predict exactly which duration will correspond to a reduction in strength, it will be necessary to perform repeated measurements until the strength is less than, for example, 50%, and to interpolate the precise duration. The figure of 50% does not have to be the same as the end-of-life criterion. Since these measurements are generally destructive, multiple specimens will be required.

- Define the size and number of specimens required for each temperature, since each temperature will require a separate oven or immersion tank. Five specimens may be necessary to define tensile strength with sufficient precision. It is more economical to expose a larger number of specimens but withdraw only two after each duration until the strength has fallen by at least 30%. Only then increase the number to five. Specimens for any particular duration should be taken from different locations in the oven or tank. In any case reserve specimens should be installed in case the durations have to be extended.

- Plot the retained strength against time and determine the rate of change. Figure 2.41 shows a typical series of measurements of reduction in strength at different temperatures, from which linear rates of change with time can be derived. The rate of degradation is then the inverse of these times. These rates provide five points for the Arrhenius graph as shown in table 2.2. Alternatively determine the times to a specific change in strength, e.g. 50% of initial strength, noting that this will generate a graph of opposite gradient.

- Examine each test sample for any change in the nature of degradation or of failure, for example, the growth of a barrier layer on the surface or circumferential cracking on the fibre surface, or increased ductility as evidenced by the geosynthetic modulus and peak strain at the higher temperatures. Light and Scanning electron microscopy is a useful aid to this purpose. If a change is observed, only those results should be retained which are regarded as being representative of long-term degradation. If process of degradation comprises two or more separate stages, these should be identified at each temperature and separate extrapolations should be made for each stage.

- Ideally start the tests at the highest and lowest temperatures first. The results of the tests at the highest temperature and shortest duration can then be used to modify the tests for the intermediate durations, in order to improve the likelihood of obtaining the required result in a time that is neither too short nor too long. The more information that can be obtained in advance, the greater is the possibility of obtaining a satisfactory Arrhenius graph within an acceptable time. The longest tests at the lowest temperature need to be started early in order not to lose any time.

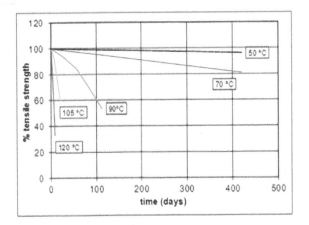

Fig. 2.41 Relation between tensile strength and time following immersion of polyester yarns in water at different temperatures.

Table 2.2 Measurements of the rate of degradation at various temperatures and derivation of the parameters for plotting on an Arrhenius diagram.

| Rate of degradation: % strength loss/day | Log (rate of degradation) | Temperature °C | Inverse absolute temperature K⁻¹ |
|---|---|---|---|
| 0.074 | -1.13077 | 120 | 0.002545 |
| 0.019 | -1.72125 | 105 | 0.002646 |
| 0.0051 | -2.29243 | 90 | 0.002755 |
| 0.00045 | -3.34679 | 70 | 0.002915 |
| 0.000093 | -4.03152 | 50 | 0.003096 |

### 2.3.9.8   Evaluation of Arrhenius tests

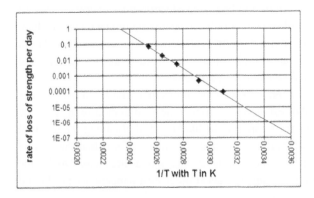

Fig.2.42 Parameters from Table 2.2 plotted as an Arrhenius diagram. The evident straight line fit demonstrates that Arrhenius' formula applies and that the relation may be used to predict the rate of degradation at lower temperatures.

On completion of the tests plot the times to a particular retained strength or other parameter against the inverse of the absolute temperature in K (figure 2.42). If Arrhenius' formula applies this plot should be a straight line. If it is curved, or if one of the points – a "rogue" point – fails to lie on the line defined by the other points, Arrhenius' formula does not apply. The reason is frequently because at the temperature of the rogue point the mechanism of degradation has changed. If the rogue point corresponds to the highest temperature, then this point should be disregarded and an additional set of tests performed at a new, lower, temperature. If the rogue point corresponds to the lowest temperature, and particularly if it indicates that the rate of degradation at that temperature is more rapid than would otherwise be expected, then any extrapolation to lower temperatures is invalid.

The rate of degradation $A$ should be expressed in absolute units (not %) per time, and the temperature $\theta$ should, as mentioned, be in K. We then set $y = \ln A$ and $x = 1/\theta$. The formula for the straight line and its gradient b can be evaluated using standard regression analysis as:

$$y = y_{mean} + b\,(x - x_{mean})$$

where $y_{mean}$ and $x_{mean}$ are the means of all values of $y$ and $x$ respectively, and

$$b = S_{xy}/S_{xx}$$

$$S_{xx} = \Sigma\,(x - x_{mean})^2$$

$$S_{xy} = \Sigma\,(x - x_{mean})(y - y_{mean}).$$

The Excel functions SLOPE and INTERCEPT can also be used. For the example above the intercept $= \ln A_0 = 12.502$ (corresponding to a theoretical rate of $10^{12.502}$ when $1/\theta = 0$) and the slope $= -E/2.3036R = -5376$, making E $= 103$ kJ/mol. By inserting the design temperature, expressed in K, into this equation one can predict the rate of loss of strength in service. In the case above the rate of loss of strength at 20 °C is deduced to be $1.42 \times 10^{-6}$ per day and the strength remaining after a lifetime of 100 years will be 94.8%.

Where the parameter being measured is not the final parameter of interest, a further correlation is required. For example in measurements of the hydrolysis of polyester fibres, where the intrinsic viscosity of the fibres was measured after exposure to hot water at different temperatures and for various durations, a separate correlation was established between intrinsic viscosity and tensile strength which enabled the residual strength of the hydrolysed fibres to be predicted (Schmidt et al 1994, Burgoyne and Merii 2007). Sangam and Rowe (2002) measured the antioxidant depletion time of HDPE geomembranes in air and water at elevated temperatures and used Arrhenius' formula to predict the time to zero oxidation induction time at various temperatures. This is the point at or just before which the mechanical properties are expected to degrade.

If the activation energy ( $E$ ) of the process is known, it may be sufficient to determine the rate of degradation at one temperature only and then to draw the Arrhenius line through that point with the slope corresponding to $E$. This procedure has been used for the oxidation of geomembranes with $E$ = approx. 60 kJ/mol. If this procedure is used, the consequences of any error in measurement or assumed value of $E$ should be assessed.

The calculation of $RF_{CH}$ is explained in chapter 3.7.

### 2.3.9.9 Uncertainty of the extrapolation

Arrhenius diagrams are frequently liable to scatter. Due to the logarithmic time scale any error in gradient can lead to large errors in the predicted time to failure. ISO 2578 states that the correlation coefficient of the Arrhenius plot for three temperatures shall be greater than 0.95.

For reinforcing geosynthetics ISO TR 20432 requires the calculation of a contributory safety factor R2 and, while this is not specified for other applications, the procedure is recommended in order to present a life prediction with sufficient confidence. The upper confidence limit of the Arrhenius line, corresponding to the maximum rate of degradation, is calculated from the formula:

$$y = y_{mean} + b \, ( x - x_{mean} ) + t_{n-2} \sigma_0 \sqrt{[1 + 1/n + ( x - x_{mean} )^2 / S_{xx} ]}$$

where, in addition to the quantities defined above,

$t_{n-2}$ is the value of Student's t corresponding to $n - 2$ degrees of freedom and the required level of probability. Suppose that the confidence level required is a 95% probability that the predicted lifetime will be less than the confidence limit shown. There is a lower confidence limit (which otherwise does not concern us: to calculate it substitute minus for plus in front of $t_{n-2}$ in the equation for $y$ above) and the probability of lying between both limits is 90%. The probability of the lifetime lying outside these limits is 10% or 0.1, which explains the use of this number in the Excel function for $t_{n-2}$ which is TINV(0.1,n-2). As a check that the right criterion has been chosen, note that as n tends to infinity Student's t tends to 1.64. $\sigma_0$ = $\sqrt{[( S_{yy} - S_{xy}^2 / S_{xx} )/( n - 2 )]}$ n = number of measurements The graph of $y$ (= ln $t$ ) and $x$ (= 1/ $\theta$ ) is a hyperbola whose centre is at the point ( $x_{mean}$, $y_{mean}$ ) and which fans outwards with increasing extrapolation (figure 2.43). The normal level of probability considered is 95% one-sided, such that one point in twenty is expected to outside the limit. The lower branch of the hyperbola would represent an lower rate of degradation which is of no interest here.

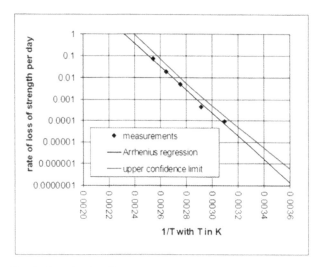

Fig. 2.43 Arrhenius diagram showing 95% upper confidence limit.

As a check,

$t_{n-2} = 2.353$

$n = 5$

$\sigma_0 = 0.1219$

Lower confidence limit:

$y = 12.515 - 5380\,x + 0.2870\sqrt{[1.2+(x - 0.002791)^2\,/(1.917 \times 10^{-7})]}$

The definition of $R_2$ is explained in section 3.7.5.4.

### 2.3.10  Changes in the state of the material: transitions

It has been clearly stated that there must be no change in the physical structure of a material over the full range of temperature and time covered not just by the accelerated tests but also by the service conditions. Transitions' are not allowed. This is a fundamental rule of accelerated testing. For example, in the hydrolysis of PVA fibres there is a transition in the rate of degradation between 40 °C and 80 °C. PVA is not hydrolysed in water, it is degraded in strongly oxidative environments. Polyvinyl alcohol (PVA) is made by hydrolyzing polyvinyl acetate. Fibres are used in geotextiles. They have a high strength and stiffness and a good resistance to alkalis (e.g. concrete), lower concentrations of acids, and oils. The Tg is in the range of 80 °C and thus needs to be considered for durability by Arrhenius extrapolations, as is the case e.g. for PET and PA. The resulting step jump in the Arrhenius diagram means that only tests at 40 °C and below may be used for life prediction of these fibres at typical service temperatures (Nishiyama et al 2006). This is achieved by sensitive measurement of the ultra-violet absorption and its correlation with the tensile strength of the fibre. In their analysis of the hydrolysis of polyesters Schmidt et al (1994) detected a slight change in gradient at 70 °C and made an evaluation for the points below 70 °C only.

There are however situations where it can be shown that the change has no effect on the process being measured and that measurements made beyond the transition are valid. In the measurement of creep of polyester geosynthetics using the stepped isothermal method (see section 3.4.2.9), the temperature is raised as high as 90 °C in spite of there being a known transition at 70 °C or below. This transition is in the physical state of the amorphous regions. However, the creep properties of the fibres are governed by the crystalline regions together with the tie molecules which cross the amorphous areas without adopting an amorphous morphology. Since neither the crystalline regions nor the tie molecules are affected by the transition, acceleration at higher temperatures is accepted and has so far proved to be valid (Thornton et al 1998, Greenwood et al 2004).

### 2.3.11  Sequential processes

The oxidation of polypropylene and polyethylene may be regarded as two or more thermally activated proces following one after the other. In this case each individual process is considered separately and a separate Arrhenius diagram is constructed to determine the lifetime of each stage. The overall predicted lifetime is the sum of these durations. More details and references are given in chapter 3.1.

### 2.3.12  Index tests

Where the degradation of geosynthetics in particular environments is sufficiently well understood, European and International standards have defined indicative or "index" tests to assure that the product should have a minimum lifetime of 5 years, 25 years and in some cases 100 years. An index test should be easy to perform, to evaluate and interprete. It should be unambiguous, should require no subjective judgment and give no optional alternatives. The test should be valid for all geosynthetics of a particular kind or of a particular polymer. The duration should be as short as possible and the test should, of course, be cheap.

The tests are based on Arrhenius or other accelerated testing and are described in more detail in subsequent chapters. The environment to which they refer is limited to a maximum temperature of 25 °C, the geosynthetic should be covered within a set time, the soil should be uncontaminated and its pH should lie between 4 and 9. These conditions are satisfied for the vast majority of applications in Europe, and in many of them the environment is even more benign. The tests are, emphatically, an assurance of a minimum lifetime and not life predictions. The geosynthetics are not expected to disintegrate suddenly on completion of this minimum, any more than food sold by a supermarket will be covered with grey mould the day after its printed sell-by date. Nevertheless, the minimum duration denotes a time after which the likelihood of failure, and the need for replacement and repair, will inevitably if inexorably increase.

In some cases manufacturers have sought to demonstrate longer lifetimes by simply extending the duration of the index tests. This is only valid where the rate of degradation is uniform (Mode 2 in figure. 2.37), such as may apply to the hydrolysis of polyester fibres but is not the case for oxidative degradation of polyolefins.

## 2.3.13  Limits of prediction

Life prediction assumes that the mechanism of degradation is sufficiently well understood. In general it assumes that there is one dominant form of degradation at any one time. Where the degradation is complex, comprising several simultaneous and interacting forms of degradation, the process of prediction will require simplifications which make the prediction less certain. Oxidation is an example of where life prediction is at the limit of its capability, due to the multiple processes acting simultaneously. Each of these will depend on the structure of the geosynthetic and the additives and impurities it contains. Since we cannot define tests for each individual formulation, the accelerated tests have to be of one general type and their interpretation somewhat empirical.

A further limit is set by the time available for testing. ISO 2578, which refers to the ageing of plastics in general, requires that the longest test (i.e. at the lowest temperature) should last at least 5000 h (about 7 months) up to the time at which the critical property is reached, and that the corresponding duration for the highest temperature should be at least 100 h (about 4 days). It also recommends that, effectively, the lowest test temperature should not be more than 25 °C above the service temperature to which the data are to be extrapolated. These numbers are useful as a guide, but are not always achievable for the long durations required for geosynthetics.

ISO 2578 also describes the use of reference materials, for example when comparing the performance of a new product with an old one whose performance is well understood. Fewer tests and wider temperature bands are then permitted. This could be of use for products such as polyethylene and polyester of similar quality, but a comparable testing system has not yet evolved for geosynthetics.

Authorities and producers alike are not prepared to wait longer than about one year for the results and consequent life predictions. This means that the temperature of testing remains uncomfortably high, typically 60 °C, while the design temperature can be 20 °C or less, a difference of 40 °C.

In such situations it may be advisable to include a cautionary factor. ISO/TR 20432 attempts to do this with the factor $R_1$ for accelerated creep tests, and to some extent with the factor $R_2$ for accelerated environmental tests. These are described further in Section 3.7.5.

# References

*Chapter 2.3*

- Burgoyne C, Merii A, On the hydrolytic stability of polyester yarns, Journal of Materials Science Vol. 42, pp2867-2878, 2007, Originally published in 1993 as a Cambridge University Engineering Department Report.
- Elsevier: Nobel Lectures, Chemistry 1901-1921, Elsevier Publishing Company, Amsterdam, 1966.
- Van 't Hoff: http://www.profburgwijk.nl/PBW/Wijk/Straatnamen/VantHoff.aspx.
- EOTA. Assumption of working life of construction products in Guidelines for Technical Approval, European Technical Approvals and Harmonised Standards. European Organisation for Technical Approvals Guidance Document 002, 1999.Greenwood J H, Trubiroha P, Schröder H F, Franke P, Hufenus R., Durability standards for geosynthetics: the tests for weathering and biological resistance Geosynthetics: Applications, Design and Construction; ed. De Groot MB , den Hoedt G, Termaat R. Balkema, Rotterdam, 1996, pp 637-642.
- Greenwood, J.H., Kempton, G .T., Watts, G.R.A. and Brady, K.C. 2004. Comparison between stepped isothermal method and long-term tests on geosynthetics. 3$^{rd}$ European Conference on Geosynthetics, Munich, Germany, 2004, ed. Floss, R., Bräu, G., Nussbaumer, M., Laackmann, K. pp 527-532.
- Nishiyama M, Yamamoto R, Hoshiro H. Long-term durability of Kuralon (PVA fibre) under alkaline condition. International Inorganic-Bonded Fibre Composites Conference (IIBCC), Sao Paolo, Brazil, 2006. pp120-133.
- Müller-Rochholz J, Bronstein Z, Schröder H F, Ryback Th, Zeynalov E B, von Maubeuge K P. Long-term behaviour of geomembrane drains – excavations on landfills after up to 12 years service. Geosynthetics – State of the art – Recent developments, eds. Delmas P, Gourc J P. Proceedings of the Seventh International Conference on Geosynthetics, Nice, France. Balkema, Lisse, Netherlands, 2002, pp565-568.
- Peggs I, Geomembrane liner durability: contributing factors and the status quo, in Geosynthetics - protecting the environment, ed. N Dixon et al, Thomas Telford, London, 2003.
- Sangam H P, Rowe R K. Effects of exposure conditions on the depletion of antioxidants from HDPE geomembranes. Canadian Geotechnical Journal, Vol 39, 2002, pp1221-1230.
- Schmidt H M, Te Pas F W T, Risseeuw P, Voskamp W., The hydrolytic stability of PET yarns under medium alkaline conditions, 5th International Conference on Geotextiles, Geomembranes and Related Products, Singapore, 1994.
- Thornton J S, Paulson J N, Sandri D. Conventional and stepped isothermal methods for characterizing long-term creep strength of polyester geogrids. Sixth International Conference on Geosynthetics, Atlanta GA, USA, 1998, pp 691-698.
- Verdu J, Colin X, Fayolle B, Audouin L Methodology of lifetime prediction in polymer aging Journal of testing and evaluation Vol 35 No 3 pp289-296.

# CONTENTS

# CHAPTER 3
# DEGRADATION MECHANISMS AND
# REDUCTION FACTORS

## 3.1    Chemical degradation and oxidation

J.H. Greenwood, H.F. Schroeder

### 3.1.1    Introduction to chemical degradation

Polymer geosynthetics degrade principally due to chemical reactions with the environment. The rate at which this occurs depends strongly on the temperature and other conditions. Before proceeding with polyolefins, a few general remarks are made:

* Many reactions occur at the polymer surface. Thus thicker materials are clearly more resistant than thinner ones.
* The mineral particles in a moist soil are surrounded by a thin layer of adsorbed water. Within this layer the concentration of active molecules or ions is higher than in the bulk fluid including, possibly, metal ions emanating from the soil particles themselves. These active molecules are formed by a reaction between a soil component and the fluid phase, for example hydroxyl ions formed by the hydrolysis of carbonates or ferric or ferrous ions formed by the hydrolysis of silicates. The molecules or ions have to move by diffusion or convection to the surface of the geosynthetic with which they are to react, and although distances are small it may be the speed of this transport rather than the rate of the chemical process itself that controls the rate of reaction. A change in temperature can lead to a change in the controlling process.
* Chemical reactions between solid reaction partners are generally slow in comparison to reactions with those involving fluid phases.
* The high surface area of fine geotextile fibres will increase the rate of chemical surface attack.
* The external reagent may be consumed at the surface. Otherwise it can permeate into the polymer and, provided that this occurs fast enough, such as with the diffusion of molecular oxygen or water, then a reaction can occur throughout the whole thickness of the material. Ionic molecules usually find it less easy to permeate. The rate of diffusion of the external reagent may be the factor that controls the rate of the whole reaction sequence.
* In semicrystalline polymers any such reactions are more likely to take place in the less ordered amorphous regions since the tightly packed crystalline regions are relatively impermeable.
* Chemical reactions can take place in the main chain of the polymer or in the side chains. Rupture of the main chain is more critical because it will directly reduce the strength of the polymer.

129

- If the type of reaction that is most likely to occur in a polymer is not already known from experience, it can be deduced by observing compounds carrying the same functional groups but with shorter chain lengths and thus lower molecular weights. These compounds are easier to handle. Such comparisons are limited by the fact that the steric effects of macromolecules, i.e. the relative movement of longer polymer chains, branches or functional groups, cannot generally be simulated by these small model molecules and have to be considered separately. Examples are the effect of the interaction between neighbouring methyl groups in isotactic polypropylene (PP) favouring oxidation and the zipper effect when eliminating hydrochloride from polyvinylchloride.

## 3.1.2   Oxidative degradation of polyolefins

Polyolefins include polyethylene (PE) and polypropylene (PP) and their copolymers and blends, all of which are commonly used in geosynthetics. They degrade principally by oxidation, which is a reaction between the polymer and molecular oxygen. The basic chemistry of this reaction is understood fairly well (Bolland and Gee 1946). However, the whole process of oxidation in solid state polymers is so complex and heterogeneous that its nature and its kinetics still leave a lot of open questions.

Polyolefin materials are mainly semi-crystalline, containing crystalline and amorphous regions. Oxygen permeates the amorphous regions of the polymer and oxidation takes place throughout these regions, both in the bulk and at the surface. Depending on the supply of oxygen, the temperature and the rate of surface reaction, the overall rate of reaction may be determined either by the rate of the oxidative chain reaction, or by the rate of diffusion and permeation of oxygen into the polymer. Particularly important is the effect of oxidation on the tie molecules that bridge amorphous regions and link the crystallites (see section 2.1.2). If these are broken, the strength of the material may be directly reduced.

Thus the molecular result of oxidation is the breakage of polymer chains, leading ultimately to a loss of strength and to an increase in the number of reactive functional groups. For polymers such as PE there may additionally be a change in the structure of the polymer such as crosslinking, leading to higher molecular mass and less flexibility.

Oxidation of polyolefins is a chain reaction whose initiation requires a source of energy for the generation of radicals which start the reaction. The chain reaction is usually structured into the following steps: chain initiation, chain propagation, chain branching and chain termination. A more comprehensive description of this reaction is given in Annex 3.1- 1.

The source of energy is most commonly thermal energy, represented by the temperature of the material, but can also be provided by ultraviolet light or other high energy radiation, energetic molecules left over from material production, mechanical shear during compounding or installation, or high permanent stresses. Transition metal ions from the environment and or catalyst residues or hydroperoxides from the material production can also assist in initiating oxidation. The result of this spike of energy is to break a bond in the polymer creating molecular fragments, known as radicals. A few of these can carry an electric charge.

In a succession of chemical reactions the radicals combine rapidly with the oxygen and then react with the polymer. These so-called propagation and branching reactions produce peroxides and other active compounds, releasing further radicals which combine with further oxygen molecules and start new chain reactions. When the local supply of oxygen dries up, or the number of active radicals is very high, termination reactions can occur which finally slow down or stop the chain reaction by reducing the number of active radicals. These reactions can produce smaller molecules containing, for example, reactive double bonds, or larger molecules with crosslinks leading to a broadening of the molecular mass distribution.

The different stages in this chain reaction and the possible side reactions proceed at different speeds (Zweifel 1998, p. 3, reactions 1.5-1.8). The rates of the reactions depend on their respective activation energies and thus on temperature. For example, the primary combination with oxygen proceeds much faster than the subsequent creation of the hydroperoxide. These reactions depend on the relevant chemical compounds being locally available. This in turn may be controlled by their appropriate rates of diffusion, in particular the speed with which oxygen can permeate on its way into and through the polymer in comparison with the rate at which it is consumed by the reactions. The more highly oriented amorphous and the crystalline regions in polyolefin materials, particularly in highly oriented fibres, may present a considerable barrier to diffusion. On the other hand a high surface-to-volume ratio such as is found in fibres and tape yarns will greatly increase the accessibility of oxygen. Under any one set of conditions the chain of reactions that make up oxidation will proceed at the speed of the slowest individual reaction.

These reactions and transport processes all depend on temperature and on external oxygen pressure, both of which can be increased to accelerate the rate of reaction. However, since each reaction can depend differently on temperature or pressure, if the temperature or pressure is raised the overall rate of reaction can change from being governed by one reaction to being governed by another. In some reactions the activation energy, which for extrapolation purposes should be independent of temperature, has been shown to decrease at lower temperatures. The cause of such a variation may be a change in the molecular reaction mechanism, or in the relative rates of chemical reactions or transport processes. This makes accelerated testing even more difficult to interpret unless performed at relatively low temperatures, when it would take many years to complete.

In the oxidation of isotactic polypropylene the steric interaction between neighbouring methyl groups plays a critical part and leads to an increase of the sensitivity to oxidation. It also leads to a rapid decrease in melt viscosity during processing, resulting in a rapid loss of molecular mass.

In contrast, the oxidation of polyethylene can generate a considerable number of double bonds. It can also cause crosslinking, resulting in an increase in molecular mass and in melt viscosity.

Small concentrations of transition metal ions can act as catalysts in the oxidation reactions in addition to the part they play in the initiation. Of these the ferric ($Fe^{3+}$) ion is the most common but copper, titanium and manganese and other ions have also been shown to be important. Some of these metal ions can originate from constituents of minerals in the soil, while others are more likely to be the residues of catalysts used to initiate the original polymerization of the monomer.

More complex reactions take place during the extrusion and drawing of the polymer into sheet or fibrous form, when the temperature can briefly reach 200 to 300 °C. The compounds left over from these reactions may influence the long-term durability. For similar reasons, high temperature ageing tests may not sufficiently simulate the reactions found during long term exposure at much lower soil temperatures.

## 3.1.3    Additives for Polyolefins

The following is a short survey of the principal additives used, comprising antioxidant stabilizers, UV-stabilizers, metal deactivators and acid scavengers. Comprehensive information is to be found in Zweifel (1998, 2000) and in Bart (2005).

### 3.1.3.1    Antioxidant stabilizers

Even though the high degree of orientation and crystallinity of polyolefins can retard oxidation, for many materials the addition of antioxidants is essential. Under normal conditions pure unstabilized PP can degrade within a year, even in the dark.

Some comments on the mode of operation of antioxidants are given in Annex 3.1-2.

The antioxidants were introduced briefly in chapter 2.1.3. Here we shall examine the main types used in polyolefins and explain their purpose. In many cases a combination of antioxidants is used: the prefabricated package is referred to as a masterbatch and is specially designed for each type of application.

Much heat is needed during the manufacturing process to enable the materials to be processed in the molten state. At these relatively high temperatures of around 200°C so-called processing antioxidants operate, notably hindered phenols and organic phosphites, designed to suit the individual processing method. On cooling the surplus antioxidants, together with the products of the reaction, remain in the polymer. In most cases these reaction products can be disregarded, but in others acid scavengers are required as described in section 3.1.3.4.

Long-term antioxidants such as aromatic amines, thioesters, sterically hindered phenols (ShPh) (figure 3.1 (left)) and sterically hindered amines (HAS) (figure 3.1 (right)) have the purpose of retarding oxidation of the geosynthetic during its service life and at service temperatures. These stabilizers either interrupt the chain of radical initiated reactions or provide an alternative means for terminating it. Initially the strong ShPh antioxidants completely suppress oxidation, resulting in an induction period during which there is no measurable change in tensile properties.

Fig. 3.1 (left) Essential functional (antioxidant) part of a sterically hindered Phenol (SHPh): The phenolic OH-Group bound to the benzene ring, which is sterically hindered by the neighboured butyl groups(+). (right) Essential functional (antioxidant) part of a sterically hindered Amine (HAS): The piperidine ring containing the amine group (N H) included and two sterical dimethyl hindrances ( < ).

HAS decrease the rate of oxidation even after the induction period created by ShPh has ceased. During the induction period the ShPh antioxidants are oxidised, i.e. they are transformed into a chemically relatively inactive species. Antioxidants such as HAS are able to regenerate for extended periods but are generally not quick and strong enough to retard oxidation to the level that would be needed to create an induction period.

The overall effectiveness of an antioxidant depends on its chemical design, its concentration and clearly on the other components of the geosynthetic polymer.

After the consumption of the antioxidants the material reacts with the radicals, which then degrade it. Some materials such as PE can deactivate radicals rather effectively for a limited period. This intrinsic stability is achieved mainly by forming crosslinks. In other materials such as PP oxidation commences immediately after consumption of the antioxidants and is accompanied by a reduction in mechanical properties. These materials therefore have a lower intrinsic oxidation stability.

Antioxidants such as HAS, if present, will therefore retard but not fully suppress oxidation, allowing for a slower reduction of mechanical strength.

### 3.1.3.2 Methods for monitoring antioxidant content

The content of antioxidants can be monitored at any stage by chemical analysis, for example ultraviolet (UV) spectroscopy, Fourier Transform Infra-red (FTIR) spectroscopy, high pressure liquid chromatography (HPLC) or gas chromatography (GC) coupled with mass spectrometry (MS) (Bart 2005). In many cases the content can be monitored either by measuring the oxidation induction time (OIT) (ISO 11357-6, ASTM D 3895, ASTM D 5885) using differential scanning calorimetry (DSC), or by using the initiated cumene oxidation test ICOT (Bart 2005, p. 47, Zeynalov 2000, Schroeder 2001, Schroeder 2002) method, which is related to long term antioxidant consumption. The limits of application of OIT have been summarised by Pauquet (1993). An OIT trace is shown schematically in figure 3.1.5. The OIT is the duration that starts with the introduction of oxygen into the system and finishes with the rise in temperature caused by the oxidation reaction.

We present now a short review of the thermo-analytical methods relevant to long term ageing. For further information refer to the literature.

Differential Scanning Calorimetry, DSC (see Ehrenstein 2004 and Riga 1997 and references for ASTM, ISO, EN) is a method of thermo-analysis that measures any kind of change in the thermal energy of materials resulting for example from chemical reactions or morphological changes such as melting. Testing is performed under controlled heating/cooling in selected atmospheres. Oxidation and crystallization are both exothermic reactions that produce energy, the first being a chemical reaction and the second a physical process. Melting is an endothermic process that consumes energy (endothermic). These and further processes can easily be detected and measured by DSC which generates a plot of heat flow against temperature (see figure 3.2). The arrows to the left of the diagram indicate the direction of endothermic and exothermic processes. The processes may be characterized by a peak as shown in figure 3.2, by their point of inflection (Glass transition temperature) as shown in figure 3.5, or by a temperature range. The area under the peak is related to the molar enthalpy of the process. Some examples of DSC are shown in figures 3.3 a to d. The figures show the initial heating trace, whose shape is determined by the material superimposed with thermal effects resulting from manufacturing. After heating the material is cooled and then heated for a second time. In this second heating these thermal effects are erased, and the trace may show additional recrystallization effects, as shown in figure 3.3 a (the smaller peak at about 90 °C, near to the melting peak at about 120 °C). Similar changes in the morphology of the specimens, such as in the degree of crystallinity, may appear following extended durations at elevated temperatures. The relatively linear DSC traces in figures. 3.3 a, b and d show that up to 85 oC, ageing proceeds normally. This means that no anomalies are to be expected when the results are extrapolated to lower temperatures. The peaks in figure 3.3 a show that for this material the material properties may cause problems if testing is performed at temperatures higher than 80 °C. Figures a and c demonstrate the need to perform a second heating together with the intermediate cooling. The effect of the cooling process depends strongly on the rate at which it occurs. The DSC-run c for a PE-blend shows anomalies in the temperature range up to 80 °C, showing that irregularities in the Arrhenius plot or changes in the material appearance might occur if oven ageing or autoclave (oxygen pressure) ageing are performed at temperatures up to 85 °C. Other thermal analysis methods include determination of loss of mass (thermogravimetric analysis or TGA), of changes in dimensions (thermomechanical analysis or TMA) or of changes in mechanical properties (dynamic mechanical analysis or DMA) related to heating and cooling in controlled atmospheres.

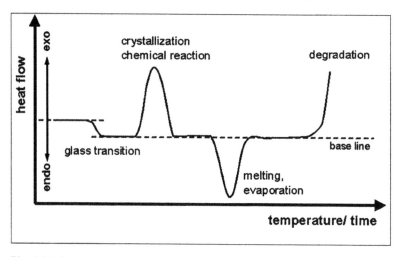

Fig. 3.2 Schematic diagram of a DSC-plot illustrating different types of thermal effects courtesy of Dr. D. Robertson, BAM 4.3).

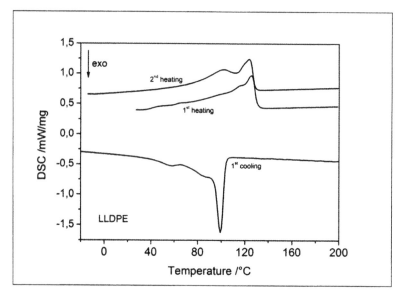

Fig. 3.3a DSC-Diagram of a LLDPE copolymer material courtesy of Dr. M. Boehning and Dr. D. Robertson, BAM 6.6 and BAM 4.3.

Information from thermal analysis should therefore be taken into consideration in planning long-term tests on geosynthetics, in particular the test temperatures. This information only relates to physical and not chemical effects since tests are performed under nitrogen.

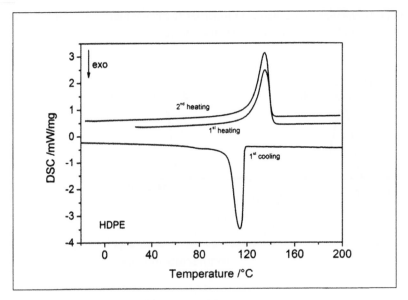

Fig. 3.3b DSC-Diagram of a HDPE material.
(courtesy of Dr. M. Boehning and Dr. D. Robertson, BAM 6.6 and BAM 4.3.)

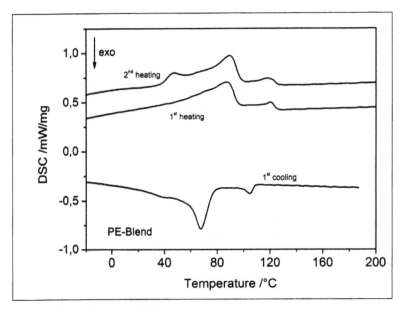

Fig. 3.3c DSC-Diagram of a PE-Blend.
(courtesy of Dr. M. Boehning and Dr. D. Robertson, BAM 6.6 and BAM 4.3.)

### 3.1.3.3 Individual antioxidants

Sterically hindered phenols (ShPh), main functional group see figure 3.1 (left), are antioxidants which interrupt the chain reaction by deactivating the radicals and are thus called primary antioxidants. They can operate very effectively both during processing and at service temperatures. Great care and experience has to be applied in their selection and formulation, in metering the correct concentration, and in ensuring their protection against UV light, since their UV stability is generally low. Their activity and compatibility with the polymer matrix is determined by their chemical structure and should be checked. Their resistance to extraction and evaporation depends mainly on their molecular mass. Antioxidants containing ester groups as shown in figure 3.4 may be hydrolyzed by inner or outer hydrolysis (see section 3.2), in which case deactivation may occur more quickly.

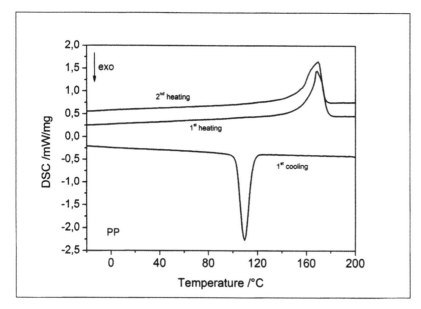

Fig. 3.3d DSC-Diagram of a PP-material.
(courtesy of Dr. M. Boehning and Dr. D. Robertson, BAM 6.6 and BAM 4.3.)

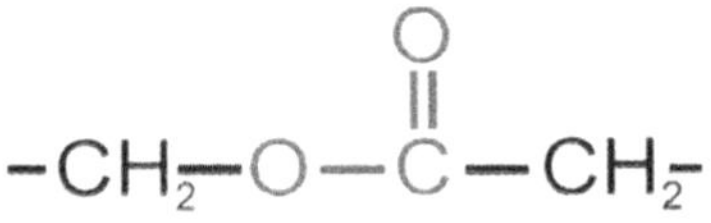

$$-CH_2-O-\overset{\displaystyle O}{\overset{\|}{C}}-CH_2-$$

Fig. 3.4 Essential functional part of an ester, the ester carboxyl group (-O-C=O). (The neighbouring methylene groups (- CH2-) are not primarily essential, but impact on the ester properties.)

Organic phosphites only operate at processing temperatures, at which they remove the hydroperoxides which act as chain initiators, and are called secondary antioxidants. These phosphites have to be chemically designed to improve their resistance to hydrolysis, which would otherwise result in phosphoric acids that could degrade the polymer or other additives.

Organic phosphites and sterically hindered phenols (ShPh) can operate synergistically. In order to maintain the effectiveness of the long-term antioxidants it is advisable to double the amount of organic phosphites (in comparison to ShPh) in the masterbatch.

Thioesters are sulphur based processing stabilizers, which while acting as an antioxidant, produce strong acids and therefore require acid scavengers. Thioethers are similar but operate at lower temperatures.

Hindered amine stabilizers (HAS) and hindered amine light stabilizers (HALS), main functional group shown in figure 3.1 (right), operate at service temperatures and are widely used to provide long-term durability. They do not operate at the high temperatures above 150 °C used in the normal OIT-Test nor in the molten state during material processing. A higher molecular weight will improve the resistance to extraction by solvents and evaporation. The properties of the amine group of HAS-stabilizers can be modified by substitution in the HAS molecule. The stabilizer has therefore to be selected to match the pH of the application, using the technical data sheet of the manufacturer.

Bromine compounds can be used as flame retardants but may meet with opposition for environmental reasons: flame retardancy is not a priority for geotextiles.

HAS are generally not appropriate for polyesters, since the amine group would introduce aminolysis of the esters. Aminolysis is a chemical process related to inner hydrolysis.

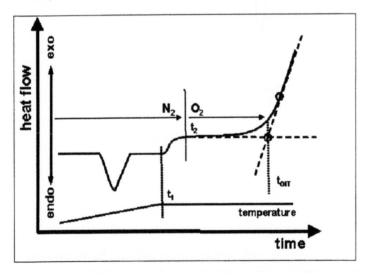

Fig. 3.5 Schematic diagram of an analysis by DSC of standard OIT at oxygen pressure of 1 bar or OIT at elevated oxygen pressure; $t_1$ time of arrival at measurement temperature; $t_2$ time of gas change from nitrogen to oxygen; tOIT extrapolated time for the start of oxidation; $t_{OIT}$ - $t_2$ = OIT (Courtesy of Dr. D. Robertson, BAM 4.3).

### 3.1.3.4   Acid scavengers

Acid scavengers are additives such as metal soaps that are introduced into the polymer formulation to neutralize acids as mentioned above. Metal soaps are metal salts of weak organic acids such as stearic acid or lactic acid. An example is calcium stearate. They also include inorganic products such as hydrotalcite or zinc oxide.

If additives contain sulphur (chemical symbol S) or phosphor (chemical symbol P) then acid scavengers should be included in the masterbatch to counteract any acid residues or products formed by their action.

### 3.1.3.5   UV-Stabilizers

The photons which make up ultraviolet light have sufficient energy to initiate photo-oxidation. Active radicals are formed, leading to the same chain reactions that follow thermally initiated oxidation. Note that these reactions can continue even in the absence of light ("dark reactions").

Ultraviolet stabilizers comprise absorbers and stabilizers. Ultraviolet absorbers, such as o-hydroxybenzophenones or carbon black, absorb ultraviolet light in competition with the reactions with chromophores contained in the polymer. Chromophores include hydroperoxides, molecules carrying ketone groups and combinations of polymer molecules with oxygen. The energy absorbed is dissipated as heat, i.e. the energy is transformed into internal vibrations and rotations of the UV-absorbing stabilizer without any further radiation.

The carbon black particles should not be so large as to initiate cracks in the polymer. On the other hand, they should be large enough to absorb the required wavelength while retaining a large surface area with the correct surface chemistry (Mwila et al. 1994). Typical particle sizes range from 25 to 75 nm. Their concentration is generally 2 to 3 % and their dispersion should be uniform. Carbon black is generally supplied as a masterbatch in a matrix resin and requires specialist preparation. For thin materials such as fibres the application of carbon black as a UV protection is clearly critical.

The other UV absorbers are mostly chemical compounds which have a high extinction coefficient in the relevant UV range (300 – 400 nm). The absorption of UV light is determined by the Lambert-Beer Law, from which the effective concentration of the stabilizer can be calculated for the thickness of the material. Their application to thin fibers is thus critical since a minimum layer thickness is required to ensure protection. In their excited state these compounds feature some phenolic properties in the same manner as phenolic antioxidants. Thus they may react with radicals and may be destroyed or consumed unless they are protected by further antioxidants (ShPh or HAS). Practical experience has shown that UV absorbers are used up after a certain UV exposure.

UV quenchers provide UV protection by a different mechanism. This group of stabilizers works by deactivating the energy transferred from UV excited chromophores to the quencher to generate heat. The transfer works by a long range mechanism. Concentrations are generally about 1% relative to the polymer matrix (Zweifel 1998, pp 63). Typical quenchers are specially masked Nickel compounds. The UV absorbance of these compounds is low in contrast to UV absorbers. Since the quenching action is independent of thickness of the polymer product, UV quenchers can be used in thin materials. The compounds also exhibit some antioxidant properties such as decomposition of hydroperoxides and scavenging radicals. Hindered amine light stabilizers (HALS, related to HAS) also operate by means of radical scavenging and quenching of excited polymer chromophores. They do not operate as UV absorbers.

### 3.1.3.6  Metal Deactivators

Transition metals, particularly those with multiple valencies such as copper, iron, vanadium, manganese and titanium, can facilitate the cleavage of hydroperoxides and thus accelerate the oxidation process by a catalytic reaction.

Metal deactivators (Zweifel 1998, pp 107) are introduced to reduce the effect of these metal ions by converting them into masked, i.e. relatively inactive, chemical compounds. Where fillers are used, it is important that if these contain critical metal ions then they should be compatible with the polymer and its additives and not catalyze oxidation. Transition metal ions arising from catalyst residues or other processing contaminants may also be present in the polymer and require the addition of metal deactivators.

### 3.1.3.7  Additives: summary

In **summary**, any polyolefin geosynthetic is likely to contain an "additive masterbatch" or package of additives to improve its resistance to oxidation including photo-oxidation (weathering).

- Selection of these requires specialist knowledge in order to avoid unexpected interactions between the additives or unwanted side effects (incompatibility). A typical masterbatch recipe might contain as a minimum:Processing antioxidants (organic phosphites, ShPh)
- Acid scavengers (metal soaps)
- Long term antioxidants such as ShPh and HAS
- Ultraviolet stabilizers (Carbon black, UV absorbers, UV quenchers, HALS)
- Metal deactivators
- For the purposes of life assessment it is important to recognize that:Without antioxidant additives polyolefins, particularly polypropylene based polymers, would have inadequate long-term durability.
- Antioxidants retard the oxidation process in various ways. In doing so they are themselves consumed, after which uninhibited oxidation can start and would lead to a quick degradation of the mechanical properties. This retardation is observed as an induction or incubation period during which the gradual depletion of antioxidants can be monitored by chemical analysis or thermo-analysis such as OIT testing. For materials stabilized with HAS the rate of loss of mechanical strength is reduced, but there is no actual induction period. Thus a combination of both types of stabilizers is recommended.
- Only certain antioxidants such as SHPh and HAS are designed to increase the long-term durability (long-term antioxidants); others such as organic phosphites are designed solely to combat oxidation during processing (processing antioxidants).
- Some antioxidants can be sensitive to ultraviolet light, such as those containing aromatic rings like SHPh. Others can be lost by evaporation, by leaching by water or other solvents at the surface, by migration into neighbouring materials or by exudation. Exudation means a segregation or blooming at the surface of plastics that is observed mainly at ambient temperatures if the solubility of the additive is exceeded. It may be caused by incompatibility with other additives or be due simply to the lower solubility when the polymer is cooled to ambient temperature.

Ester bonds present in some antioxidants, e.g. in Irganox 1010, may be cleaved, if rather slowly, by hydrolysis at high humidity or by long term contact with highly alkaline aqueous media. Leaching, evaporation and exudation are particularly relevant in materials with a high surface-to-volume ratio such as fibres. It is the responsibility of the material designer to assure that , in addition to having good chemical durability and good compatibility with the resin, the stabilizers used are sufficiently resistant to leaching for the application in hand, having for example a sufficiently low solubility in water especially where there is a continuous flow of fresh liquid.

Neither the user nor the manufacturer of a geosynthetic is likely to know the exact additive package in the material. This section is intended to show that specialists are available who have methods for determining the content, state and efficacy of the additives.

## 3.1.4    Accelerated testing of polyolefins

### 3.1.4.1    General remarks

The lifetime of existing polyolefin geosynthetics under ambient conditions in soil is usually so long that up to now there has been little experience of long-term changes in properties. This knowledge will no doubt increase with time. In most applications the geotextile will be exposed permanently to the environment, even if this environment may change with time. This rules out any opportunity for speeding up the timescale. In addition, geosynthetics once installed are difficult to access or inspect, let alone do they allow for samples to be removed.

Life prediction therefore requires accelerated testing. We have already seen that oxidation is a complex sequence of processes, each of which can proceed at a different rate, with each rate depending on temperature in a different way. That was even without the additives. The additives retard oxidation and control the rate of degradation, but in doing so are consumed. This allows oxidation proper to commence, followed by a reduction in tensile properties which for many applications is the critical end-of-life parameter.

In theory it would be possible to construct a model of the oxidation and retardation procedure and to establish its dependence on temperature and oxygen concentration, and then to perform detailed measurements to determine the many governing parameters for each material and environment. In reality this requires accurate knowledge of the polymer and its additives, together with numerous precise and regular measurements and their evaluation. This can only take place in a research laboratory, and the assessments would be very complex and even so be valid only for the assumed model of degradation model. This is not a realistic way to tackle the issue.

The first and simplest method for predicting lifetime was to treat the entire oven oxidation and ageing history as one simple thermally activated process.

Samples of the polypropylene textiles used in the Ooster Schelde barrier in the Netherlands were regularly extracted and subsequently subjected to oven ageing tests, allowing their performance to be compared with that of virgin material and the accelerated ageing to be scaled against real life performance (Wisse et al. 1982, 1990). The degradation of polypropylene fibres in a circulating or gravity driven air oven (i.e. an oven without forced circulation) can be remarkably sudden – fibre one day, black powder the next – which means at least that the end point can be determined with sufficient precision.

As a result of this extensive work a number of problems were noted:
- The range of test temperatures is limited. The upper test temperature must remain well below the melting temperature of the polymer (80°C is a guide given e.g. by Achimsky (1997), Gugumus (1999)), while the lower test temperature has to be high enough to obtain results within a reasonable duration. Further information on the selection of valid

testing temperatures is provided by thermoanalytical methods such as DSC (see section 3.1.3.1).

- The lowest test temperature is more than the recommended maximum of 25 °C (see section 2.3.9.7) above the service temperature (e.g. 45 °C for 20 °C service temperature). With this level of extrapolation the predicted life should be subject to a safety factor.

- The mechanism of oxidation, and the efficacy of the antioxidants, can differ at higher temperatures. Wisse et al. (1982, 1990) noted that at typical test temperatures above 80 °C the activation energy for PP oxidation lies between 65 and 110 kJ/mol, but that for service temperatures below 50 °C the activation energy could be 50-55 kJ/mol or even lower (see also Ding et al. 2001). This has the serious consequence that lifetimes extrapolated from high temperature tests can be too long, i.e. optimistic.

- Surface cracking can increase the access of oxygen which could invalidate any prediction based on uncracked material; conversely at elevated temperatures these cracks may even heal. Surface cracking has been observed for PP oxidation under wet conditions (Salman et al. 1998 (wet and dry conditions), Schroeder et al. 1997, Richaud et al. 2007). It is probably caused by rapid leaching of antioxidants at the specimen surface. Yet it is difficult to visualize that PP material would be converted into a powder within 24 h (Wisse et al. 1990) without first developing cracks both at the surface and then across the entire thickness. The possible distribution of these cracks and their consequences are still under discussion. Surfactants are common and, when they are present, environmental surface cracking may accelerate this further.

- Tests in air may be untypical since the concentration of oxygen in water is about one-seventh of that in air at ambient conditions (D'Ans Lax 1967). Within the soil the oxygen concentration is likely to be lower, depending on the rate of oxygen transport and of oxygen consumption, and at most equal to that of normal air (Richter J. 1986). In the absence of further information, the oxygen content of normal air (partial pressure 0.21 bar) is used as a conservative basis for assessments of durability.

- The temperature dependence of the solubility of oxygen in polymers is not considered in oven testing, leading to doubts concerning its validity. Generally this process is exothermic and thus solubility decreases with increasing temperature. Thus though the oven ageing index test according to EN ISO 13438, methods A and B, is still valid, extrapolations of temperatures using Arrhenius' formula may not be conservative due to changes oxygen content. It is also possible that at such high temperatures the polymer may be subject to irreversible structural changes due to the temperature alone. Additional tests including DSC could be used to answer these questions.

- The water present in soil can, apart from dissolving in and swelling the geosynthetic, lead to leaching of some additives (e.g. antioxidants) and possibly to hydrolysis of additives of the polymer. This leaching can have a major effect on changes in additive concentrations and is influenced strongly by the surface-to-volume ratio, the speed, direction, composition and flow of soil water and the rate of oxidation under these conditions. It is also significant in the temporarily anaerobic regions of landfills.

- Transition metals such as iron accelerate the oxidation of polyolefins (see 3.1.2, 3.1.3.6). Their effect has not yet been considered in the screening tests. This remains a task for future research, since iron, its oxides and hydroxides are ubiquitous in soil.

ISO 13438 Methods A and B use a simple oven ageing test as a screening test for the resistance to oxidation of polyethylene and polypropylene based geosynthetics, with the test durations doubled for reinforcing applications. EN 14575 defines a similar test for geosynthetic barriers running at markedly lower temperatures. These tests are based on existing results and practical experience on some materials over at least 20 years. It must be emphasized that they are just screening tests to ensure a minimum lifetime of 25 years under set environmental conditions and with the purpose of eliminating poor quality material. They are not life assessment tests: It is quite possible that a material which passes the index test could survive for 100 years or more under favourable conditions. The test does not cover the leaching of stabilizers, for which there is a separate standard for geomembranes (EN 14415). The test does not consider the effect of oxygen diffusion which may limit the rate of oxidation in a geosynthetic and can restrict the oxidation reaction to a purely surface reaction. Note the particular danger that this effect can cause poor quality materials to pass the screening test without any problems if they possess a sufficiently thick cross-section.

Very recently WG 5 of CEN TC 189 proposed to use the oven test only in a screening test for service life > 50 years, based on the method described in ISO 13438, due to lack of sufficient and clear test results of the autoclave tests performed to date. Further experience and research is needed to fully comprehend this test.

At the same time WG 5 of CEN TC 189 proposed to add a leaching procedure prior to the execution of the oven test.

It is therefore not possible to make an accurate life prediction based on a single screening test. For example, when two thick PP split film yarns were tested at high temperatures to EN 13438 Methods A1 and A2, one stabilized and the other unstabilized, they both passed the screening test, since the rate of oxidation of the unstabilized yarn was limited by oxygen diffusion and not by the level of stabilization (Schroeder 2001). While this is certainly not representative of their long-term behaviour, EN ISO 13438, Method B, could not serve its purpose of preventing their passing for 25 years' usage.

Arrhenius tests based on simple oven ageing tests have been used, so far rather successfully. These oven test models do not take into account the complexity of natural ageing, the sequential nature of oxidation in stabilized polyolefins nor the loss of stabilizers by other mechanisms. Thus great care is necessary to avoid misleading results.

The current situation regarding tests for long-term resistance of geosynthetics to oxidation is still controversial. Details of the methods for testing and extrapolation are not defined with sufficient precision. The requirements will differ according to application, since the relevant chemical and physical ageing mechanisms and their interactions are not known precisely.

Progress is being made in life assessment but this is still a subject for discussion. Some of the procedures used are described in the following sections.

### 3.1.4.2 Oven testing

Simple oven testing is performed according to EN ISO 13438:2004, methods A and B in ovens without forced air circulation, as already mentioned. Because of its relative convenience it is a widely used method, but its limitations should not be ignored. The limits of the method are emphasized in the introduction and annex A of the standard and in section 3.1.4 of this document. The details of the ovens to be used are described in ISO 188: 2007.

For screening lifetimes of at least 25 years, test conditions are defined as follows: Specimens of PP are exposed to heated air at 110 °C for 14 days, or 28 days for reinforcement applications. For PE based materials the temperature is 100 °C and the durations are 28 days and 56 days respectively. After these exposures the retained strength should be at least 50% of its initial value.

The high testing temperatures may lead to problems by inducing structural changes and may even cause melting of some flexible PE-blends with low melting temperatures (see figures 3.1.3a and c). For products with thick cross-sections such as extruded geogrids or coarse tape yarns oxidation at high temperatures may be oxygen diffusion limited, especially if poor quality material tested, which is not representative of oxidation at lower temperatures. This will even happen during oven testing after the antioxidants have been consumed. This could lead to predictions of too long lifetimes.

The oven ageing method may be performed at lower temperatures in line with EN ISO 13438:2004, methods A and B, to generate an Arrhenius plot, provided that one can accept test durations of four years or longer. The very dry conditions in the oven exposure and the absence of leaching conditions may not be representative of the environment in the application envisaged. During oven ageing the loss of additives, which is often the main reason for the premature failure of materials, is only possible by evaporation. Leaching of the additives in water could be modelled by additionally immersing the specimen in water before oven ageing, as was performed in the work of Wisse et al 1982. This pre-leaching would reduce oven testing durations.

The general rules of Arrhenius testing (section 2.3.9.7) should be followed. The highest testing temperature should not exceed about 85 °C for polyolefins (Achimsky 1997, Gugumus 1999), as used in the index test for the resistance of geosynthetic barriers to oxidation, EN 14575.

This results in very long testing durations. To help with evaluation of the results additional tests should be performed to detect small changes in the content of antioxidants which precede the fall in strength. This is described in more detail in the three phase model below (section 3.1.4.3).

A problem will occur if at these elevated temperatures the Arrhenius plot still turns out to be nonlinear. Lower temperatures have then to be used. The problem could be related to the solubility of oxygen as described in 3.1.4 and be corrected if the dependence on temperature is known.

For geosynthetic barriers a different oven ageing method, EN 14575:2005, is applied in Europe for screening for a durability of 25 years and for comparing barrier products. The oven has gravity air circulation. The test can be performed on PE, PP and bitumen based materials, provided that they are resistant to the testing temperature which is fixed at 85 °C. The exposure duration is 90 days. A DSC plot will provide valuable information on the temperature performance for critical materials, see figure 3.3c. If the material is not resistant at the testing temperature of 85 °C, screening could be performed by choosing an appropriate lower temperature and prolonging the test duration according to Van't Hoffs rule (see section 2.3.9.6). This would be analogous to the procedure chosen for PP and PE in EN ISO 13438, method A and B.

### 3.1.4.3   Three stage Arrhenius testing

Meanwhile in the USA (Hsuan and Koerner 1998, 2005) it was recognized that many antioxidants result in a three stage oxidation process (figure 3.6), which was defined as follows:

A Depletion of antioxidants (Stage A).
In this initial incubation phase antioxidants are depleted by several simultaneous processes including consumption by oxidation and physical losses such as extraction, evaporation and exudation, the total effect of which can be monitored by OIT. For waste disposals the loss of stabilizer due to oxidation is generally low compared with the loss by physical processes.

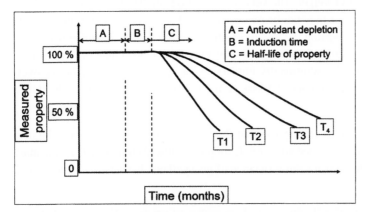

Fig. 3.6 Three stage model of Hsuan & Koerner 1998 for oxidation of HDPE geomembranes (courtesy of Prof. Dr. G. Hsuan and Prof. R. M. Koerner).

The mechanical properties are unchanged since the changes in molecular mass are limited and do not reach the critical mass limit for this to occur. The end of this phase is defined as the point at which the OIT reaches zero or a predefined minimum level such as e.g. 0.5 minutes, indicating zero antioxidants (Bartholomeo 2004, Rowe and Rimal 2009).

B Induction time (stage B).
These low OIT values form the start of stage B. There remains no change in the mechanical properties, in spite of the lack of protection by antioxidants. However, a further induction phase is created by the intrinsic oxidation stability of the material and the initial lag phase resulting from the kinetics of the oxidative chain reaction mechanism. The changes in molecular mass are limited and have not yet reached the level necessary for a reduction in strength, thanks to the ability of the material to deactivate reactive radicals such as by crosslinking in PE. This second stage may be quite long where the conditions are favourable. The end of this stage is defined by the point at which the mechanical strength starts to decrease, showing that the critical molecular mass has been reached. Seen chemically, the concentration of reactive intermediates has reached the level at which rapid chain propagation and branching can proceed.

C Polymer Degradation (Stage C).
The mechanical strength decreases and there is a significant reduction in molecular mass. The material is no longer capable of deactivating reactive radicals. Uninhibited oxidation and rapid oxygen consumption take place, tie molecules are destroyed and the mechanical properties change accordingly. Stage C ends when the strength reaches a pre-defined end point, such as 50% of its initial value.

In this procedure samples were exposed in devices specially designed to simulate a landfill at four temperatures ranging from 55 to 85 °C (Hsuan and Koerner 1998, 2005) and regular specimens.

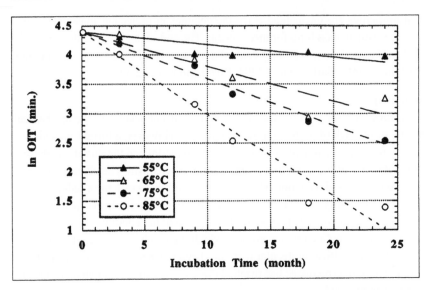

Fig. 3.7 ln OIT for exposure durations at specified temperatures under conditions given in Hsuan and Koerner 1998 for lifetime evaluation of HDPE geomembranes. The slopes of the straight lines give the OIT depletion rate needed for figure 3.8 (with courtesy of Prof. Y.G. Hsuan and Prof. R.M. Koerner).

Extractions were made. Samples were tested both for mechanical properties and OIT. The duration of each stage is extrapolated by Arrhenius plots to the assumed service temperature of 20 °C, the end-of-life criterion for the third stage being a 50% reduction in strength. An Arrhenius relation is set up for each of the three stages independently and used to predict the duration of each stage at the service temperature. These three durations are finally added to give the predicted life.

As an example the diagram of Hsuan and Koerner for the evaluation of the duration of stage A is given in figure 3.7 and the Arrhenius diagram is shown in figure 3.8.

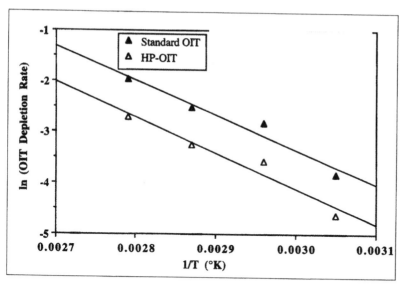

Fig. 3.8 Arrhenius Diagram ln OIT of the antioxidant depletion rate S over inverse absolute Temperature (with courtesy of Prof.Y. G. Hsuan and Prof. R.M. Koerner).

During ageing the additives may be consumed by chemical reaction with oxygen, or lost by physical processes including evaporation, leaching, exudation, or migration at the surface as explained. The change in antioxidant content of the material may be monitored by oxygen induction time measurements according to ASTM D 3895-98 or ISO 11357-6:2002(E). These are performed on small samples (about 5 mg) at high temperatures (e.g. 200 °C). Thus the material is molten and the effectiveness of the antioxidants will differ from that at the service temperature. A typical OIT measurement curve has been shown in figure 3.5. High pressure OIT (HPOIT) according to ASTM D 5885:2006 operates at a lower temperature (e.g. 150 °C) to approximate more closely to service conditions and applies oxygen at high pressure to compensate for the reduction in acceleration. It requires special testing equipment and longer testing durations.

The depletion rate at the service temperature can be derived by extrapolating figure 3.8, from which the duration of Stage A can be predicted. The diagram shows the plot for Standard OIT (upper line) and HPOIT results.

The three stage model is attractive since it takes into account the observed sequence of different types of processes, even though these remain a mixture of individual processes. It applies particularly to HDPE and has been used widely to predict the life of geomembranes. Hsuan and Koerner (1998, 2005) provide examples of this three stage analysis. The total life prediction is 712 years at 20 °C reducing to 109 years at 40 °C, with Stage 3 providing over half the predicted lifetimes for the material tested. This separation into stages can in principle be used with other lifetime assessment methods. The second phase will be less pronounced for PP based materials, since their intrinsic stability is generally low. The third phase may be short for ShPh stabilized materials for which there is a rapid fall-off in strength. The method is less appropriate in the case of HAS stabilization but might be modified for this situation, thus testing is nevertheless recommended.

Both OIT methods have the limitation that they are performed at very high temperatures and they do not differentiate between processing antioxidants such as organic phosphites and long-term antioxidants such as ShPh. OIT tests do not therefore provide the information we need on long-term stability directly, but they are an important tool for monitoring the state of antioxidants during oxidation of geosynthetics.

Stabilizer content can be measured by means of direct chemical analysis or by ICOT, which uses the initiated oxidation of cumene for the integral determination of the long-term activity of antioxidants (Schroeder and Zeynalov 2002, Barth 2005). ICOT is not able to determine the content of specific antioxidants, but only the overall content of each long-term SHPh and HAS type antioxidant. It thus provides an attractive and less expensive alternative to chemical analysis. It may be appropriate to apply several methods in order to minimize the operating expense.

### 3.1.4.4   Autoclave testing

Oven ageing of polyolefins either takes place at temperatures that are rather high, or it takes too long. At these high temperatures the mechanism of oxidation may not be the same as at ambient temperatures. The rate of oxidation is likely to be dominated by oxygen diffusion so that the dimensions of the specimens may be more significant than the stabilization, as shown by the example that EN ISO 13438 method B could not differentiate between a thick stabilized split film PP yarn and one which was unstabilized (see 3.1.4.1). An alternative method was sought. In addition to using lower temperatures and taking less time than, say, two years for assessments of service lifetime, the objectives were to have a single test for screening oxidative durability for as many different materials as possible and to include a liquid environment to allow for leaching, which is often the main cause of loss of additives.

Fig. 3.9 Autoclave testing facility with data monitoring equipment in the background
(courtesy of Dr. M. Boehning, BAM 6.6).

Pressurized air and oxygen have been used for many years to accelerate the ageing of elastomers and other polymers. For applications where oxidation is so fast that it is limited by oxygen diffusion these methods are in fact not the right choice. However, under in-soil conditions the lower temperatures lead to very low rates of oxidation, so that oxygen can reach an equilibrium concentration over the whole product width by diffusion. Thus the rate of oxidation should not be limited by the rate of diffusion and can be accelerated by increasing the oxygen concentration. Here, autoclave testing becomes the method of choice. For safety and handling reasons the testing volumes were kept relatively small. In the method developed by the Federal Institute for Materials Research and Testing (BAM) in Berlin (Schroeder 2000) the pressure vessel is large enough to contain geosynthetic samples immersed in a standard lightly alkaline solution, which is intended to favour leaching of phenolic antioxidants. If appropriate, other solutions can be used to represent different environmental conditions. Because of its size, volume about 8 liters, and the 51 bar pressure the walls have to be made sufficiently thick, and a pulley mechanism is needed for the lab assistant to lift the lid and also if necessary the vessel (see figure 3.9). The equipment has proved sufficiently handy and very reliable for more than ten years. An interlaboratory trial between three well-known institutes, KIWA-TBU, SKZ and BAM is in progress and has so far delivered good results that will be published in the near future.

Particular advantages of this method compared with simple oven ageing are:
- The combination of enhanced oxygen pressure, elevated temperature and leaching of antioxidants by an appropriate aqueous medium provide a highly accelerated test, leading to shorter testing durations.

- Liquid immersion and stirring assure a constant temperature throughout the whole vessel.
- A liquid medium provides a closer approximation to the soil environment, and allows for leaching of stabilizers and reaction products. The composition of the liquid medium may be adapted to special environmental conditions or to special additive (e.g. antioxidant) types. The intensity of leaching should be controlled by selecting an appropriate liquid medium and the intervals for its renewal.
- Elevated oxygen pressure reduces the likelihood of oxygen diffusion being the controlling factor for oxidation at elevated testing temperatures.
- Liquid immersion reduces strongly the risk of ignition or explosion of fine fibres or of certain surface preparations in dry oxygen. (This risk in dry oxygen may be checked in advance by using smaller specially designed autoclaves.)
- It is easily possible to perform the test under nitrogen in order to eliminate or identify slow thermal or physical effects in the specimens that are not related to oxidation, such as swelling, shrinkage and recrystallisation.
- The vessel and equipment are easy to clean. The test results are independent of the quality of the air.

Disadvantages are that the liquid environment cannot simulate pure evaporation of stabilizers from the polymer surface. If for a particular application oxygen diffusion is more likely to be the limiting factor, leading to surface oxidation, then an oven test is more appropriate. Furthermore there are difficulties with testing foamed materials, materials that are sensitive to elevated pressure or decompression or those sensitive to aqueous solutions. Thus each test has its place.

For evaluation of the test results the Arrhenius equation has to be modified by an oxygen pressure term, representing the impact of oxygen concentration on the oxidation reaction. The modified Arrhenius function currently used is not optimal, yielding results which are thought to be conservative. Work is in progress to resolve this.

Müller et al.2003, Richaud et al. 2008 point out that under autoclave test conditions there may be a significant direct reaction between phenolic antioxidants and oxygen. This is relevant to all oxidation tests. If it is shown to be significant an appropriate correction factor may have to be introduced.

Vink and Fontijn (2000) pointed out that the dependence of the rate of oxidation on pressure may be nonlinear or reach an upper limit.

ISO 13438 Method C uses the oxygen pressure test as a screening test with set parameters of 51 bar (i.e. 50 bar above atmospheric pressure) and 80 °C, and a test duration of 14 days for non-reinforcing materials. The pass criterion is 50% retained strength. As described above, these are minimum values to assure a lifetime of 25 years under set conditions and are not life predictions. For reinforcing applications the test duration is 28 days. Geomembranes such as liners for tunnels (Boehning 2011, Boehning 2008, EAG-EDT 2005, ZTV-ING 2007), though not yet explicitly covered by the standard, can be tested by the same method and using the same parameters.

To use the method for life prediction it is necessary to perform the test over a range of pressures and temperatures (figure 3.10 and figure 3.11). For execution of an Arrhenius test evaluation a pressure term has to be added. The dependence on pressure and temperature is then shown as a three-dimensional diagram which is extrapolated to service temperatures (figure 3.12) (Schroeder 2008).

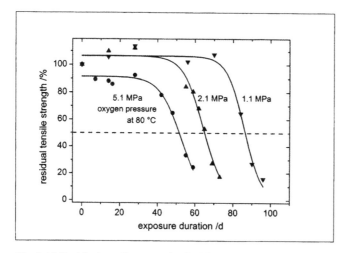

Fig. 3.10 Residual tensile strength of a PE geomembrane (tunnel liner) after exposure to different oxygen pressures (courtesy of Dr. M. Boehning and Dr. D. Robertson, BAM 6.6 and BAM 4.3).

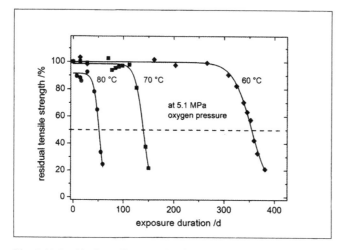

Fig. 3.11 Residual tensile strength of a PE geomembrane (tunnel liner) after exposure at different temperatures (courtesy of Dr. M. Boehning and Dr. D. Robertson, BAM 6.6 and BAM 4.3).

Mathematically, we amend the Arrhenius equation, (see section 2.3.9.6). The equation already includes the pre-exponential factor $A_0$, the apparent activation energy $E$, the time to a set end point criterion $Y$ and exposure temperature T in the form,

$1/ Y = A_0 \exp ( E /R \, T)$ (Arrhenius equation).

We add an extra term to take account of oxygen pressure $P$:

$1/ Y = A_0 \exp [( E + CP )/R \, T]$

where $C$ is a constant coefficient. We then plot $\ln (1/ Y)$ against $P$ and $1/T$ (T in K) to give a three-dimensional diagram as shown in figure 3.12. This can be used to predict the value of $Y$ at the service temperature (e.g. 25 °C = 298 K) and pressure (e.g oxygen partial pressure in air 0.21 bar). The value of $Y$ extrapolated to the service conditions for the geosynthetic is denoted as the *lifetime parameter* $\Theta$. This is illustrated in the plots in figure 3.12 and figure 3.13, the first with $1/ Y$ plotted logarithmically and the second plotted linearly.

Note that this evaluation has been performed with TableCurve 3D (Systat software, Inc., Richmond, USA) as a regression analysis of the following equation:

$z = A + B \, x + C \, xy$

In which the variables are set as $x = 1/T$, $y = P$ and $z = \ln ( 1/ Y )$.

For these evaluations values of $Y$ were derived from figure 3.10 and figure 3.11 which represent the original measurements of the autoclave test for this life assessment. Curves were fitted using a standard function. The values $Y$ are taken from the intersections of the fitted curves with the dotted line representing the end point criterion of 50% retained strength.

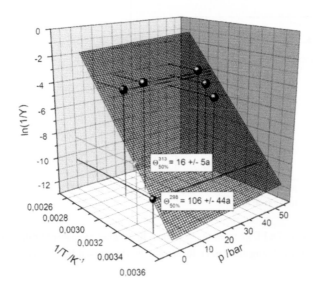

Fig. 3.12 Three-dimensional diagram showing the extrapolation to ambient oxygen pressure of test results at elevated temperature and pressure for two service temperatures ($1/Y$ in logarithmic coordinates). The point $\Theta^{313}$ shows the lifetime parameter for 40 °C (313 K) to be $16 \pm 5$ years; $\Theta^{298}$ shows the lifetime parameter for 25 °C (298 K) to be $106 \pm 44$ years (courtesy of Dr. Martin Böhning, (BAM 6.6)).

The effect of measurement variability on the extrapolated life time parameter $\Theta$ can be assessed by introducing the extreme measured values of strength (strength $\pm$ standard deviation) into the respective plots of figure 3.10 and figure 3.11, fitting new curves to generate modified values of $Y$, and finally performing the regression procedure with these modified values of $Y$ to obtain modified values of $\Theta$. The standard deviation of these modified values of $\Theta$ is calculated and shown in the 3D plots figure 3.12 and figure 3.13.

The precision of the test could be improved by taking more measurements. The dependence on pressure depends only on the measurements at 80 °C. This is clearly not enough. Further measurements should be made at 70 °C, 21 and 11 bars, noting that measurements at the lower pressures at 60 °C would lead to excessively long test durations.

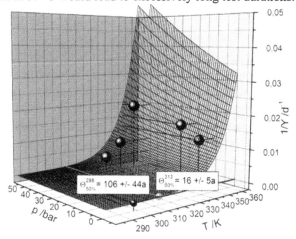

Fig. 3.13 Results of autoclave tests in general accordance with ISO 13438 method C for assessing the lifetime parameter for a tunnel liner, and extrapolated to two service conditions as in figure 3.12. ($1/Y$ in linear coordinates) (courtesy of Dr. M. Boehning BAM 6.6).

Further research is being done to resolve some open questions concerning the execution and evaluation of the test. Recently (Bartholomeo 2004) it was established that the extraction of phenolic stabilizers by the liquid medium during the autoclave test is independent of the pressure on the system. This validates the use of tests performed at the same temperature but different at oxygen pressures. The rate of extraction of antioxidants can be compared with that at normal ambient pressure.

The lifetime parameter $\Theta$ still suffers from the uncertainty of extrapolating over a temperature range of more than 25 K (see section 2.3.9.7), all the more so for a complex oxidation process when compared with the relatively straightforward inner hydrolysis of PET described in section 3.2. There is also the question of the validity at service temperatures of an activation energy derived from tests at elevated temperatures (Ding 2001; see also section 3.1.4.1).

The Hsuan-Koerner three stage approach (section 3.1.4.3) could be applied to the autoclave test by measuring OIT or ICOT antioxidant equivalents of the extracted specimens during the incubation phase to separate phase 1, antioxidant depletion, from phase 2, induction time.

If HAS stabilizers are present a clear separation would be difficult, since this stabilizer only reduces but does not completely arrest degradation of the mechanical properties. Furthermore, OIT will only approach but not equal zero (see Rowe 2009).

Identification of the start of the third phase needs additional extractions of specimens in order to interpolate the beginning of the degradation of tensile properties. The Arrhenius function modified for oxygen pressure could be applied to each of the phases. The lifetime would then be the sum of the lifetimes of the three phases, and its variances would be calculated from the three standard deviations of the results of the phases. Clearly the effort required for this method is considerably greater and thus results must show if it is cost-effective.

At the time of writing the authors are not aware of this procedure having been performed. Finance would be necessary for such a project.

A similar oxygen pressure test, see figure 3.14, has been developed in the USA, working under pure oxygen in much smaller autoclaves of about 0,5 l volume (Li and Hsuan 2004) without a liquid phase. Substantial acceleration of the oxidation has been observed here too. Specifically aimed at geomembranes, the test does not require immersion of the geosynthetic in an aqueous liquid, and thus does not take into account the loss of stabilizers by leaching. Leaching therefore has to be assessed by additional tests.

Fig. 3.14 Pressure cell with tape yarn samples for incubation at GRI (courtesy of Prof. Y.G. Hsuan).

A further application of elevated oxygen pressure is its use in HPOIT measurements, which is standardized as ASTM D 5885. By this means the test can be performed at lower temperatures and the antioxidant activity of HAS stabilizers can be detected.

The oxygen pressure methods are promising. They allow for careful control of conditions, the temperatures are not extreme, they have shorter test durations and the first (BAM) method includes directly the leaching effect of liquids. The equipment is however more expensive than for an oven, not so much the autoclave itself as the instrumentation required to control and measure pressure and temperature. At the time of writing most testing is performed in Germany and the USA alone, with some interesting research running in France (Bartholomeo 2004, Richaud 2007).

What is absent so far is a clear validation of all the oxidation test methods by comparison with real long-term durability. An assessment of their reproducibility by means of interlaboratory trials would internationally contribute to their acceptance. As mentioned above, an intercomparison test is currently running between the three German testing institutes that can perform the autoclave test.

Further comments have been made at the start of this section.

Thus both methods will continue to be used, each in its right place. For the purpose of index tests ISO 13438 currently regards them as alternatives.The challenge is to define the limits of application of each method with respect to both the type of material and the environment. Work on this is in progress.

### 3.1.5 Degradation of polyolefinic geosynthetic barriers – General remarks

In this section some general remarks will be made on geosynthetic barrier materials which, although they do not belong to the scope of this book, illustrate the problems with predicting oxidation. For further reading Schweitzer (2007) is recommended.

Other important geosynthetic materials are polyvinyl alcohol (PVA), polyvinyl chloride (PVC) and polystyrene (= polyvinyl benzene) (PS), all of which possess rather similar basic chemical structures, see figure 3.15. All have a saturated hydrocarbon backbone like PE or PP but with different substituents. Thus PP has methyl (-CH3)-sidegroups, PVA has hydroxyl (-OH) sidegroups, polystyrene has benzyl-sidegroups and polyvinyl chloride has chloride sidegroups.

Geosynthetic barrier materials, PVA, PS and PVC will be described briefly in the following sections. The hydrolysis of PET is described in section 3.2.

ISO/TS 13434, section 5.2.3, states that flexible polypropylene copolymer (fPP) geosynthetic barriers are subject to oxidation. The material has the advantage that it can be welded easily. It can be affected by several solvents such as halogenated aliphatic hydrocarbons, aromatics as well as by aliphatic hydrocarbons, organic acids, volatile organics, oils and waxes. These are only likely to occur in landfills, liquid waste containers and contaminated soil.

Chlorosulfonated polyethylene CSPE and the terpolymer ethylene-propylene-diene-elastomer (EPDM) geosynthetic barriers, are elastomeric, i.e. the resin is cross-linked. They are also prone to oxidation, CSPE less than EPDM and much less than fPP, since EPDM has a certain intrinsic stability due to its ability to forming crosslinks. The composition of EPDM and thus its properties can vary widely. Elastomers can be affected by industrial waste liquids containing high concentration of aromatic and chlorinated organic hydrocarbons. Welding of elastomers is more difficult but they can also be jointed using vertical attachments.

Modified bituminous barriers (T. McNally 2011) are also stabilized with antioxidants. They should not be exposed to non polar, aromatic, aliphatic or halogenated hydrocarbon solvents for long periods of time, nor to acidic solutions with pH < 2 and alkaline solutions with pH > 9. While screening tests exist for these conditions (EN 14414, EN 14415) for PET and polyamide geotextiles life prediction would require a full range of accelerated immersion tests appropriate to the material, the environment and the expected mode of degradation.

### 3.1.5.1   Degradation of PVA

The basic chemical structure of PVA is shown in figure 3.15, left.

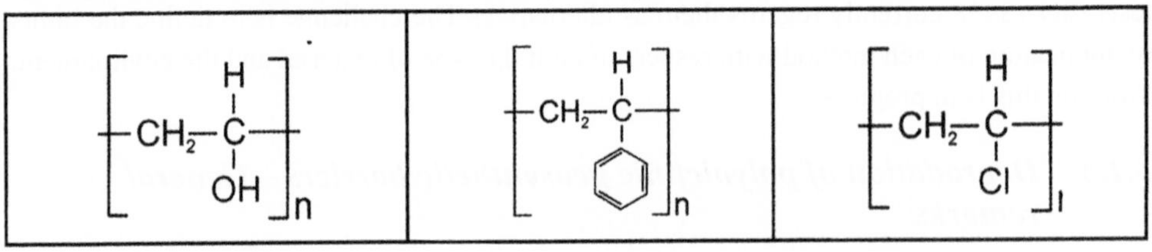

Fig. 3.15 Chemical basis structure of PVA (left), PS (centre) and PVC (right).

The polymer is obtained by hydrolysis of its acetate ester poly(vinyl acetate). For high performance PVA fibers a degree of hydrolysis greater than 98% and a sufficiently high molecular mass are essential (Nait-Ali et al. 2009). Further tacticity would increase the performance (Sakurada et al. 2007). For the stability of PVA fibers in aqueous environments the highly crystalline morphology is of the utmost importance. Amorphous PVA is water soluble in case it has a low degree of hydrolysis or if it is not acetalyzed. It is very difficult to obtain a PVA solution in water after acetalyzation. The degradation by microorganisms depends on the arrangement and the structure of the polymer. It is very difficult to obtain a PVA solution in water after acetalyzation. The degradation by microorganisms depends on the arrangement and the structure of the polymer. The development of high tenacity fibers with a very low solubility in water at temperatures as high as 80 °C has depended on the ability to produce highly oriented, highly crystalline material.

Since PVA fibers and yarns have up to now been used principally as a reinforcement for concrete, studies have been concerned mainly with their degradation in alkaline environments. The mechanism of degradation (Nishiyama et al. 2006) is oxidation, which forms carbonyl groups next to double bonds (-CO-C=CH-) in the main chain. These cause discoloration, which can be measured by UV-absorption at 235 nm.

The exposure temperatures were 20, 30, 40, 50, 60 and 70 °C and the tests at 20 °C lasted for up to 30000 h (3.5 years). The authors demonstrate that for testing durations up to 1.2 years at temperatures ≥ 50 °C the discoloration correlates with the retained tensile strength, and is thus closely associated with the scission of tie chains in the amorphous regions. They do not propose a mechanism to explain this. At lower temperatures the changes in tensile strength were too small to be measured, but by using this empirical correlation observed at higher temperatures they were able to measure very precisely the rather small changes of absorption and to correlate these absorption changes with retained tensile strength. Their assessment done on the basis of an Arrhenius plot for the temperatures 20, 30 and 40 °C results in the prediction of a residual tensile strength of 99.8% after immersion in cement extract with pH 12.6 at 20 °C for 100 years.

What is important for life prediction is that there is a step change in between 40 and 50 °C so that tests performed at higher temperatures cannot be used to predict lifetimes at service temperatures of 40 °C and below, and that the changes at the lower temperatures are so slight that very precise detection methods (here UV absorption) have to be used to assist with the prediction. The reason for the step change at the molecular scale is not explained, but could be the glass transition temperature which was not explicitly stated by the author.

Akers et al. (1989) and Hirai et al (1993) are earlier papers which are in agreement with the results of Nishiyama et al. but only at higher temperatures (above 50 °C).

The situation under very acid conditions (pH ≤ 3) is somewhat different. While under alkaline conditions the morphology is stable, at low pH the protons are able to permeate slowly, to loosen the hydrogen bond structure and thus to make it more vulnerable to oxidation. The resistance to oxidation under acid conditions (pH = 3) could be screened in line with EN ISO 13438 , method C2, for 56 d at 70 °C and 51 bar oxygen. After exposure the retained tensile strength and strain both exceeded 50% of their initial values. It was concluded that there was sufficient resistance to acids for at least 25 years. Longer term resistance could be demonstrated in the same way as reported for alkaline resistance, since a similar discoloration was observed by the author of this section.

### 3.1.5.2 Degradation of polystyrene (PS)

The basic chemical basic structure of PS is shown in figure 3.15 (centre). While oxidation of polystyrene (PS) is similar in nature to that of polypropylene, the resistance to oxidation is markedly greater thanks to the benzene ring which attracts electrons away from the tertiary hydrogen positioned on the same carbon atom as the substituent benzene. But, as in polypropylene, oxidation eventually leads to hydroperoxide residues which increase the sensitivity to further oxidation. Thus the material has to be appropriately stabilized with antioxidants.

Polystyrene requires UV stabilizers to protect it from UV light, since again the hydro-peroxide residues and other ubiquitous chromophores increase its sensitivity.

Long term stability can be assessed as for other polyolefins. The autoclave test cannot be easily used for a foamed material, since the oxygen would permeate the material during exposure and destroy it on extraction by sudden expansion (decompression). This could be avoided by lowering the pressure very slowly to atmospheric pressure at room temperature, but this take too long. Thus the oven test has to be applied, selecting temperatures appropriate to the application of the Arrhenius method (section 2.3.9.7) and possibly with a pre-leaching exposure beforehand. The duration of leaching should be adapted to the intended application. The Van't Hoff rule can be applied if no specific information on the activation energy of leaching is available. Leaching and oxidation can be handled using the Hsuan-Koerner three stage approach described in section 3.1.4.3.

### 3.1.5.3   Degradation of plasticized polyvinylchloride (PVC-P)

The basic chemical structure of PVC is shown in figure 3.15 (right). In dealing with degradation it must be realized that PVC-P generally contains about 35 % of plasticizers. These plasticizers may be a mixture of compounds, all differing in their chemistry and molecular mass which strongly influences leaching resistance. The plasticisers may also be prone to oxidation, hydrolysis and aggression of microorganisms. To achieve a plasticizing effect in PVC only the amount exceeding 15% is effective. Thus a 20% loss of plasticizer, or 15 % residual content, are generally taken as the end of life of PVC-P barriers.Loss of plasticizers leads to shrinkage of the material, depending on the plasticizer content by volume. If the recipe is confidential, the testing institute is faced with a rather complicated scenario.

While oxidation of polystyrene (PS) is similar in nature to that of polypropylene, the resistance to oxidation of PVC-P is markedly greater thanks to the chlorine atom which attracts electrons away from the tertiary hydrogen positioned on the same carbon atom as the substituent chlorine. But, as in polypropylene, oxidation eventually leads to hydroperoxide residues which increase the sensitivity to further oxidation. Thus the material has to be appropriately stabilized with antioxidants.

In planning the test procedure the thermal behaviour should first be investigated by DSC. Generally, due to the tendency of chlorine to split off as hydrochloric acid, temperatures higher than 70 °C should not be used. Appropriate aqueous solutions should be used for leaching and they should be renewed at the latest when they become saturated. If leaching takes place on one side of the specimen only, then this should be taken into account in the evaluation. The intensity of leaching depends on the composition and incident flow of the liquid. Leaching conditions have to be adapted appropriately to the conditions in the intended application.

At the time of writing there is no European or international regulation for PVA, PS and PVC-P geosynthetics, although national regulations exist such as for tunnel liners in the

NEAT test specifications in Switzerland, the German DGGT EAG-EDT(2005) and the German ZTV-ING, part 5 Tunnel (2007). These will not be commented on here.

**Summary**

- Polymer geosynthetics degrade principally due to photochemical, chemical and in a few cases microbiological reactions, together with the physical effect of the environment. The rate at which these occur depends strongly on the temperature and other conditions.

- Many effects occur at the polymer surface. Thus thicker materials are clearly more resistant than thinner ones. If the external reagent is not consumed at the surface it will usually diffuse into the polymer and its rate of diffusion may be the factor that controls the rate of the whole reaction sequence. A purely surface effect forms one extreme, while the at other extreme a the bulk-reaction takes place uniformly across the full cross-section. In reality there is mostly a mixture of both. Photochemical and microbiological degradation and leaching of additives are examples of surface effects, while under normal conditions in soil hydrolysis and oxidation are examples of bulk reactions.

- Chemical reactions between solid reaction partners are generally slow in comparison to reactions with those involving fluid phases, including thin fluid layers adsorbed on the surface.

- Reactions which rupture the main chain of the polymer are more critical since this will directly reduce the strength.

- Polyolefins, including polyethylene (PE), polypropylene (PP) and their copolymers and blends, degrade principally by oxidation. This is a complex process involving a chain of individual reactions, each of which may proceed at its own rate and have its own sensitivity to temperature. Oxidation is initiated by a source of energy such as high temperature or UV light. Transition metals can act as catalysts.

- Additives are essential to the long-term durability of polyolefins, particularly polypropylene and its copolymers and blends. Typical additives include antioxi-dants which operate at the high temperatures during processing, acid scavengers, long term anti-oxidants, ultraviolet stabilizers including carbon black, and metal deactivators. Of the long-term antioxidants sterically hindered phenols (ShPh) result in a long incubation time followed by a rapid loss of strength, hindered amine stabilizers (HAS) in a more gradual loss of strength with no incubation time.

- There are currently two principal accelerated methods for determining resistance to oxidation: oven ageing and autoclave ageing under oxygen pressure. Both are adapted as index tests in ISO 13438. Both can be used as the basis for Arrhenius tests to predict long-term durability. Both use loss of tensile strength as the end-of-life criterion for geotextiles. Autoclave ageing includes the impact of loss of additives by leaching.

- Oven ageing in air requires equipment that is simple and generally available. For geotextiles the method requires either long durations or high temperatures, and allows for the loss of additives by evaporation but not by leaching. At these temperatures the mechanism of degradation may differ, for example the rate of oxidation may be controlled by oxygen diffusion or solubility. Care has to be taken that under the testing conditions no purely surface reaction takes place, which does not represent in-soil conditions. This may be determined by microscopic and chemical analysis of the specimen crossection.

Tests performed over a range of temperatures can be extrapolated to service conditions, provided that the dependence of rate of oxidation on temperature is established.

- Thermal analysis tests carried out before long-term testing provide a simple method of determining the temperature limit beyond which nonlinear Arrhenius-plots are to be expected, caused by physical effects such as changes of morphology.

- By measuring the stabilizer content at intervals during exposure of the samples it is possible to divide the process of degradation into three stages: antioxidant consumption, induction period and loss of mechanical properties by oxidation. Arrhenius diagrams are set up to predict the lifetime for each stage, the sum of which is the predicted lifetime for the material. The method has to be modified for products stabilized with HAS where the rate of loss if strength is less abrupt.

- Autoclave ageing is promising and delivers a precious advantage in lower testing time expenses. Clearly it requires equipment that is somewhat more expensive but not prohibitively so. Shorter durations and lower temperatures can be used. Immersion in liquid allows for the loss of additives by leaching but not by pure evaporation. The American method, which is aimed more at geomembranes, is performed in dry oxygen and pre-leaching is done in special devices. Tests performed over a range of temperatures and oxygen pressures can be extrapolated to service conditions, provided that the dependence of rate of oxidation on temperature and oxygen pressure is established and further developed.

- The lifetimes of correctly stabilized polyolefins under normal soil conditions have been shown to be long compared with practical experience to date. For this reason it has not yet been possible to validate either method against long term lifetimes in soil.

- It is clear that each method may be more suitable for particular types of material and environment. Pure oven ageing, including the screening test ISO 13438 Methods A and B, may be more appropriate for materials where leaching is not an issue and for materials sensitive to elevated pressures or decompression. It is not appropriate for materials where there is pure surface oxidation, such as materials with a low resistance to oxidation. This has to be decided by experiment. For the time being it is not easily possible from a scientific standpoint to specify which method to use. Speed of testing, the inclusion of leaching and the avoidance of surface oxidation all favour the autoclave test, but the equipment for oven ageing, if applicable, is more generally available. For index testing according to ISO 13438 this standard currently treats them as alternatives.

- High tenacity PVA fibres degrade very slowly by oxidation in alkaline environments. UV absorption has been used successfully as a method for determining their rate of degradation. Polystyrene degrades by oxidation. The principal cause of degradation in plasticized PVC is loss of plasticizer.

Annexes:
*Annex 3.1-1 Schematic diagram of the oxidation of polyolefins*

(Simplified for people not familiar with chemistry)
Some preliminary comments for better understanding: Polyolefins such as PE or PP consist of hydrogen(H) and carbon(C). The whole family of such compounds is thus called hydro-

carbons. Their basic polymer molecules can be represented by the symbol RH, where R represents the whole hydrocarbon polymer molecule apart from the hydrogen H. The molecule can contain Carbon-Carbon double bonds. If a hydrogen atom is abstracted from the hydrocarbon molecule the remaining part is called a hydrocarbon-radical (abbreviated to R*, the * symbolizing the reactive radicals). Radicals are very reactive molecules. The charge of the molecule is not changed by transforming it into a radical. O represents oxygen, H represents hydrogen. The darts → represent the chemical transformation (reaction) taking place.

The following is a strongly simplified scheme of the radical initiated **oxidative chain reaction**:

**Chain initiation:**
**RH + energy (e.g. heat, radiation or shear) → R\* or in words:**
**hydrocarbon + energy → hydrocarbon-radical**

**Chain propagation:**
Very rapid reaction of the hydrocarbon radical with oxygen
**R\* + O2 → ROO\* or in words:**
**hydrocarbon-radical + oxygen → hydrocarbon peroxy radical**

And regeneration of the hydrocarbon radical by **hydrogen abstraction** from a further hydrocarbon molecule RH
**ROO\* + RH → ROOH + R\* or in words:**
**hydrocarbon peroxy-radical + hydrocarbon → hydrocarbon hydroperoxide + hydrocarbon-radical**

Thus an endless chain reaction will take place, producing lots of hydroperoxides unless the hydrocarbon radical is deactivated (see chain termination). This chain, including branching, see below, is also called an **auto-oxidative cycle**. The rate of the reaction at first is rather slow in this **induction phase**. It strongly increases with the amount of hydroperoxides formed. Further chain branching takes place at elevated temperatures or in the presence of catalyzers (e.g. heavy metal ions), since hydroperoxides are rather stable at ambient conditions.

The hydrogen abstraction just described is the rate limiting step in the autoxidative cycle provided that sufficient amounts of oxygen are available, otherwise oxygen supply becomes the rate limiting factor. Instead of RH other molecular species containing e.g. double bonds or bound oxygen may participate in the chain propagation reaction (side reactions), leading e.g. to crosslinks or carbonyl (C=O) containing species.

**Chain branching:**
The rate of oxidation will be multiplied if chain branching can take place, since many more active radicals are formed rapidly and are able to start further chain reactions.

ROOH + energy and/ or heavy metal ions → RO* + OH* or in words:

hydrocarbon hydroperoxide + energy and/ or heavy metal ions → hydrocarbon oxy-radical + hydroxyl-radical

and / or
ROOH + ROOH + energy and/ or heavy metal ions → RO* + ROO* + H2O, in words:

hydrocarbon hydroperoxide + hydrocarbon hydroperoxide + energy and/ or heavy metal ions → hydrocarbon oxy-radical + hydrocarbon peroxy-radical + water

**Chain termination:**
If many radicals are present they can meet and react and so the number of radicals will decrease. This leads to a finally reduced rate of oxidation:

R* + ROO*→ ROOR or in words:

hydrocarbon-radical + hydrocarbon peroxy-radical → dihydrocarbon peroxide

and / or

R* + R* → R-R, in words:

hydrocarbon-radicalC* + hydrocarbon-radicalC* → hydrocarbon-hydrocarbon

and / or

R* + RO* →R-O-R, in words:

hydrocarbon-radicalC* + hydrocarbonoxy-radicalCO* → hydrocarbon-O- hydrocarbon (a Dihydrocarbon ether)

and /or

2 R* → RH + Olefin, in words:

hydrocarbon-radicalC*+ hydrocarbon-radicalC* → hydrocarbon + olefin

(Comment: Olefin is a hydrocarbon containing double bonds. These compounds are more reactive than hydrocarbons containing no double bonds (alkanes).)
All molecular species formed may again participate in the chain reaction.

Other possibilities of chain termination are provided by the action of antioxidants.

## *Annex 3.1-2 Schematic diagrams of the Mechanisms of Antioxidants*

Primary Antioxidants are able to deactivate radicals. Secondary antioxidants are only able to deactivate hydroperoxides. Some simple examples are given. Further study of the technical literature is recommended (see e.g. Zweifel 1998).

**Sterically hindered phenols, aromatic amines etc. are acting as hydrogen-donors:**

This type of antioxidant compound (abbreviated to InH for the complete inhibitor including a reactive specially bound H, H for hydrogen) deactivates radicals by donation of hydrogen:

**InH + R\* (or RO\* or ROO\*) → In\* + RH (or ROH or ROOH)**

In\* is made rather inactive by special chemical design (i.e. kind of substituents) of InH in order to avoid a direct reaction with oxygen. It cannot participate in the auto-oxidative cycle, but can often be further transformed into other species also capable of deactivating additional radicals.

**Hydroperoxide Decomposers (HD):**

These compounds, such as organic phosphites or phosphonites or compounds of sulfur (prefix: thio) such as thiocarbamates or dithiophosphates are designed to reduce hydroperoxides to comparatively stable alcohols. The HD is oxidized. One example:

**organic phosphite + ROOH → organic phosphate +ROH, in words:**

**organic phosphite + hydrocarbon hydroperoxide → organic phosphate + alcohol**

The organic phosphite compounds offered on the market differ in their hydrolytic stability. Hydrolysis leads mostly to the formation of phenols and phosphoric acids. The compounds function up to processing temperatures of 300 °C; their long term performance at ambient temperatures is insignificant. The acids have to be deactivated with acid scavengers (section 3.1.3.4).

**Sterically hindered Amines:**

Hindered amine stabilizers, abbreviated to HAS, are relatively large oligomeric compounds with a very low tendency to evaporate. The compounds contain active amine functions, see figure 3.1 (right), which are each a part of a cyclic ring compound. The amine at first has to be oxidized to nitroxyl radicals, which are effective antioxidants against thermal degradation. Thus HAS stabilizers can also be bought in the more expensive active oxidized form.

Radicals of the auto-oxidative chain are deactivated by the nitroxyl radicals to low activity compounds, and the nitroxyl radical is regenerated until the ring is destroyed by some competing reaction.

The stabilizer is not active at the high temperatures under processing conditions and first needs activation by oxidation to form the nitroxyl- compound, which is the active antioxidant.

# References

*Chapter 3.1*

- Bolland, J.L. and Gee, 1946, G., Trans. Faraday Soc., 42, pp 236.
- Wisse J. D. M., Broos C. J. M., Boels W. H., 1990, Evaluation of the life expectancy of polypropylene geotextiles used in bottom protection structures around the Ooster Schelde strom surge barrier. Geotextiles, Geomembranes and Related Products, ed. Den Hoedt G. Balkema, Rotterdam, pp697-702).
- L. Achimsky, L. Audouin, J. Verdu, J. Rychly & L. Matisova-Rychla, 1997, J., Polym. Degrad. and Stability 58, pp 283-289.
- H.F. Schröder, H. Bahr und E. Lorenz, 1997 „Zur Bedeutung der Oxidation von Polyolefinwerkstoffen im Erdbau", Geotechnik, Sonderheft, Deutsche Gesellschaft für Geotechnik, pp 167-180.
- Y.G. Hsuan and R.M. Koerner, June 1998, J. of Geotechnical and Geoenvironmental Engineering, pp 532-541.
- Gugumus F., 1999, Effect of temperature on lifetime of stabilized and unstabilized PP-films, Polymer Degradation and Stability 63, pp 41-52.
- Wisse and Birkenfeld, 1982, The Long-term Thermo-Oxidative Stability of Polypropylene Geotextiles in the Oosterschelde Project, Second International Conference on Geotextiles, Las Vegas, USA, Session 8A, pp 283 -288.
- Pauquet J.R., Todesco R.V., Drake W.O., Limitations and applications of oxidative induction time (OIT) to quality control of polyolefins, Presented at the 42nd International Wire & Cable Symposium, November 15-18, 1993.
- Ding S., Khare A., Ling M.T.K., Sandford C., Woo, L., 2001, Polymer durability estimates based on apparent activation energies for thermal oxidative degradation, Thermochimica acta, 367-368, pp 107 -112.
- Mwila J., Miraftab M. and Horrocks A.R. 1994. Effect of carbon black on the oxidation of polyolefins-An overview. Polymer Degradation and Stability, 44(3), pp 351-356.
- McNally T. (Editor), "Polymer modified bitumen, Properties and Characterisation", Woodhead Publishing, Oxford Cambridge Philadelphia New Delhi, 2011, pp 264.
- Schweitzer P.A., 2007, Corrosion Engineering Handbook, Second Edition, Corrosion of Polymers and Elastomers, CRC Press, Boca Raton, London, New York.
- Salman A, Elias V. and Dimillio A.,1998, The Effect of Oxygen Pressure, Temperature and Manufacturing Processes on Laboratory Degradation of Polypropylene Geosynthetics,Sixth international Conference on Geosynthetics, pp 683-690.
- Richaud R., Farcas F., Divet L. and Benneton J.P., 2007, accelerated ageing of polypropylene geotextiles, the effect of temperature, oxygen pressure and aqueous media on fibers- methological aspects, pp71-81.
- Richter J.,1986, Der Boden als Reaktor: Modelle für Prozesse im Boden. F. Enke Verlag D'Ans Lax, 1967, 3. Auflage, Taschenbuch für Chemiker und Physiker, Band 1, p. 1205.
- Bartholomeo P., Astruc A., Massieu E., Barberis N., S. Lavaud, Benneton J.P., March 2004, Thermo-oxidative ageing of polypropylene geosynthetics monitored by thermal

analysis and mechanical tensile test, Eurogeo 3: Geotechnical Engineering with Geosynthetics, Munich Germany, Proceedings pp 415-420.

- Richaud E., Farcas F., Fayolle B., Audouin L., Verdu. J., Accelerated Aeging of Polypropylene Stabilized by Phenolic Antioxidants Under High Oxygen Pressure, Journal of Applied Polymer Science, 110, pp. 3313-3321.

- Rowe R.K., Rimal S, Sangam H.; (2009), Ageing of HDPE geomembrane exposed to air, water and leachate at different temperatures, Geotextiles and Geomembranes, 27,pp. 137-151.

- Schröder, H.F. & Bahr, H. & Herrmann, P. & Kneip, G. & Lorenz, E. & Schmuecking, I. 2000. Testing the oxidative durability of polyolefine geosynthetics under elevated oxygen pressure in aqueous liquids. Proceedings of the Second European Geosynthetic Conference, Bologna, Volume 1: 459 pp.

- H.F. Schroeder, E. Zeynalov, B. Ladeur, H. Bahr, P. Herrmann, G. Kneip, E. Lorenz and I. Schmücking, "A new Method for Testing the Oxidative Durability of Geotextiles and Geosynthetics made from Polyolefins", 11th International Techtextil Symposium for Technical Textiles, Nonwovens and Textile-Reinforced Materials, Frankfurt am Main, 23. -26.04.2001, Paper 436, Tagungs CD-ROM Messe Frankfurt 2001.

- H.F. Schroeder, E.B. Zeynalov et al. Polymers & Polymers Composites 10 (2002), pp 73.

- Zeynalov, E.B., Schroeder, H.F., Bahr, H., 2000, „Determination of Phenolic Antioxidant Stabilizers in PP and HDPE by means of an oxidative Model Reaction", Proceedings of the 6th International Plastics Additives and Modifiers Conference – Addcon World 2000, Basel.

- Zeynalov, E.B., Bahr, H., Rybak, Th., 2001, "A quantitative study of sterically hindered phenol and amine stabilizers in PP materials, Proceedings of the 7th International Plastics Additives and Modifiers Conference – Addcon World 2001, Berlin.

- Müller W., Buettgenbach B., Jakob I., Mann H., 2003, Comparison of the oxidative resistance of various polyolefin geotextiles, Geotextiles and Geomembranes, 21, pp. 289-315.

- Müller W.W., editor, Certification Guidelines for Plastic Geomembranes Used to Line Landfills and Contaminated Sites, Landfill Engineering Laboratory, BAM, Berlin.

- Robertson D., Boehning M., Schroeder H.F., Brummermann K , Kaundinya I. and Saathoff F., 2011Möglichkeiten der Abschätzung der Langzeitbeständigkeit von Kunststoffdichtungsbahnen(KDB) auf Polyolefinbasis für den Tunnelbau, FS KGEO, 03 März 2011, 12. Informations- und Vortragsveranstaltung über Kunststoffe in der Geotechnik, Tagungsband 2011.

- Boehning M., Robertson D., Schroeder H.F. EuroGeo4, Edinburgh, 2008, paper 143

- Schroeder H.F., Munz M., Böhning M. Polymers & Polymer Composites 16 (2008), 71 -79. - Li M. and Hsuan Y.G. 2004, Temperature and pressure effects on the degradation of polypropylene tape yarns-depletion of antioxidants, Geotextiles and Geomembranes 22, pp 511-530.

- Hsuan Y.G. and Li M. 2005, Temperature and pressure effects on the oxidation of high-density polyethylene geogrids, Geotextiles and Geomembranes 23, pp 55-75.
- Yoshitaka N., Nakano T., Okamoto Y., Gotoh Y., Nagura M., 2001, Properties of highly syndiodactic poly(vinyl alcohol), Polymer, 42(24), pp 9679-9686.
- Nishiyama M., Yamamoto R., Hoshiro H., Long –term durability of Kuralon (PVA fiber) in alkaline condition, 2006, 10th Int. Inorganic Bonded Fiber Composites Conference, November 15-18, Sao Paulo-Brazil, pp120-134.
- Lewin M. (Editor), 2007, Handbook of Fiber Chemistry, Third Edition, chapter 4, Sakurada I., Okaya T., Vinyl Fibers, pp 262-329.
- Nait-Ali K. L., Freitag N., 2009, Renforcements a base de fibres de polyalcool de vinyle: effet des cycles humidification/sechage et consequences sur le dimensionnement dans le cadre des norms francaises; Rencontres Géosyntiques.
- Zweifel H., 1998, Stabilization of Polymeric Materials, Springer Verlag Berlin Heidelberg New York.
- Zweifel H., 2000, Plastics Additives Handbook, 5 th Edition, Hansa Publishers, Munich.
- Bart J.C.J. Additives in Polymers Industrial Analyses and Applications John Wiley & Sons Ltd, 2005.
- ISO 188:2007, Rubber, vulcanized or thermoplastic. Accelerated ageing and heat resistance tests, section 4.1 and 4.1.3.
- ZTV-ING, Teil 5 Tunnel, BASt 12/2007, Verkehrsblatt Verlag.
- EAG-EDT „Empfehlungen zu Dichtungssystemen im Tunnelbau" DGGT 2005, VGE Verlag, Essen.
- Vink, P., Fontijn, H. F. N., 2000. Testing the resistance to oxidation of polypropylene geotextiles at enhanced oxygen pressures. Geotextiles and Geomembranes, Vol. 18, 333-343.
- Greenwood J H, Curson A. Life prediction of the oxidation of geogrids by three different methods, Geotextiles and Geomembranes (in press).

*References (books) for TA (e.g. for DSC and OIT:*
- Riga A.T., Patterson G.H., Editors; Oxidative behavior of materials by thermal analytical techniques, 1997, American Society for Testing and Materials, West Conshohocken, PA.
- Ehrenstein G.W., Riedel G. and Trawiel P., Thermal Analysis of Plastics- Theory and Praxis, 2004, Hanser Publishers Munich.

*References (standards) for TA (e.g. for DSC and OIT):*
- ISO 11357-1:2009 Plastics-Differential scanning calorimetry(DSC), Part 1: General principles.
- ISO 11357-2:1999 Plastics-Differential scanning calorimetry(DSC), Part 2: Determination of glass transition temperature.
- ISO 11357-3:2011 Plastics-Differential scanning calorimetry(DSC), Part 3: Determination of melting and crystallization temperature and of melting and crystallization enthalpy.

- ISO 11357-4:2005 Plastics-Differential scanning calorimetry(DSC), Part 4: Determination of specific heat capacity.
- ISO 11357-5:1999 Plastics-Differential scanning calorimetry(DSC), Part 5: Determination of characteristic reaction – curve temperatures and times, enthalpy of reaction and degree of conversion.
- ISO 11357-6:2008 Plastics-Differential scanning calorimetry(DSC), Part 6: Determination of oxidation induction time ( isothermal OIT) and oxidation induction temperature ( dynamic OIT).
- ISO/DIS 11357-8:2001 Plastics-Differential scanning calorimetry(DSC), Part 8: Determination of amount of absorbed water.
- ASTM D 3895: 2007 Standard Test Method for Oxidation Induction Time of Polyolefins by Differential Scanning Calorimetry.
- ASTM D 5885a: 2011 Standard Test Method for Oxidation Induction Time of Polyolefin Geosynthetics by High Pressure Differential Scanning Calorimetry.
- BS EN 728: 1997 Plastics Piping and ducting systems- Polyolefin pipes and fittings- Determination of oxidation induction time.

## 3.2    Hydrolysis: polyesters, polyamides, aramids

J.H. Greenwood. H. F. Schroeder

### *3.2.1  Introduction to degradation of polyesters*

In comparison with the complex mechanism of the oxidation of polyolefins the hydrolysis of polyethylene terephthalate (PET) geosynthetics, generally referred to as polyesters, is more straightforward. In contrast to polyolefins polyester materials have a high intrinsic durability. Thus leaching, evaporation etc. are not a problem for Polyester. Polyester fibres and bars are used in nonwoven and woven geotextiles and in geogrids, notably for reinforcement applications. In the automotive industry polyester fibres are used in large quantities as tyre reinforcements, so that for reasons of price geotextiles have adopted the same fibres for high quality reinforcing products. While this has prevented the development of fibres tailored to the geotextiles market, it has assured on the other hand that the polyester fibres used for this kind of application are generally all of a similar specification. Fibres and raw materials with a different specification may be used in nonwovens. The only polyester geotextiles currently available that are not based on fibres are laid geogrids manufactured from welded PET bars. The specifications of the raw materials for welded PET geogrids differ also in general from the PET raw material used for fibres.

Polyester fibres degrade by two independent mechanisms of hydrolysis. In acid or neutral environments, including water vapour, the dominant hydrolysis reaction takes place throughout the cross-section of the PET-products (figure 3.16), leading to progressive rupture of the polymer chains and consequent loss of strength of the fibres and bars. This reaction which is referred to as "internal" hydrolysis, is related to the number of carboxyl end groups present at the end of polymer chains (left side of PET in figure 3.16). The aggressive ions involved are hydrogen ions (protons), which can be provided from the free carboxyl endgroups (Ravens et al. 1961) of the PET-molecules or other acidic components present, like e.g. phenols from antioxidants. A higher molecular weight indicates a lower population of carboxyl end groups and thus a lower tendency to hydrolysis. In addition, the high degree of orientation and crystallinity in polyester products reduces the free volume of the amorphous phase where the reaction takes place and thus increases the resistance to hydrolysis. Thus also the morphology and possibly other additives and chemical modifications, e.g. copolymers (different diacids, dialcohols), modified carboxyl endgroups and residual content of catalysers might have an impact on resistance to hydrolysis and cannot be excluded for an adequate assessment of resistance to hydrolysis of arbitrary PET geoproducts.

| n (HOOC- ø - COOH) + n (HOCH2CH2OH)   H[-OOC-ø-COOCH2CH2-]n OH +( 2n-1) H2O | | | |
|---|---|---|---|
| terephthalic acid | ethylene glycol | PET | water |
| a dicarbonic acid | a dialcohol | a polyester | water |
| with ø = benzene ring | | | |

Fig. 3.16 Synthesis( →)/Hydrolysis( ←) of polyethylene terephthalate (PET).

In alkaline environments the more aggressive hydroxyl ions (OH- ions) cannot penetrate the bulk volume of the PET-product and degradation takes place at the surface. It is therefore known as "external" hydrolysis. Material is removed from the surface leading to a loss in strength proportional to the reduction in cross-section. Additives play a minor part. There may be some accelerating effect of residual impurities on the fibre surface causing less ordered structures like TiO2 particles or other fillers or residues of catalyzers. A supposed example for this situation is shown in figure 3.17.

In both cases the backbone of the polymer chain is attacked, but the location of attack differs from that in internal hydrolysis. This results in a lower molecular weight too. Both mechanisms are thermally activated and therefore react faster at higher temperatures according to Arrhenius' formula, but both proceed very slowly at normal ambient temperature and normal ionic concentrations (like pH, pCa).

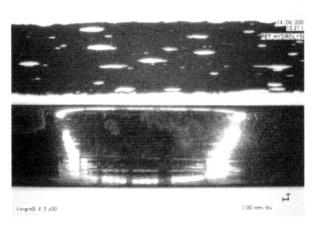

Fig. 3.17 PET-strand (8 mm x 0,8 mm)before (below) and after (above) 4 year immersion in saturated lime solution at 50 oC (courtesy of Prof. Dr. J. Müller-Rochholz).

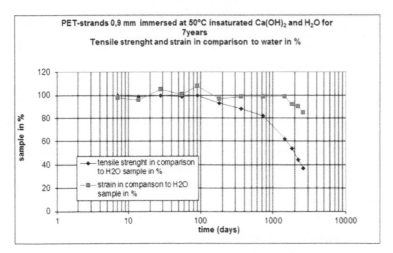

Fig. 3.18 PET strands immersed in water (red line) and in saturated Ca(OH)₂ (blue line) solution (courtesy of Prof. Dr. J. Müller-Rochholz).

Other reactions can occur at higher temperatures. Methylene groups within the polymer chain may oxidize, but more critical are the ether groups originating from the diethylenglycol content of the dialcohol component of PET. The sensitivity to oxidation can be reduced by controlling the structure of the polyester, limiting the concentration of diethylenglycol groups (Ludewig 1975) by appropriate conditions and catalyzers during synthesis. Some crosslinking can also occur under harsh processing conditions.

Resistance to internal hydrolysis is generally provided by careful structuring of the polymer rather than by the use of additives and particularly by ensuring that the polyester remains very dry at every stage of its processing. However, hydrolysis stabilizers, mostly derivates of carbodiimides, may be introduced. Such compounds reduce the number of effective carboxyl end groups which provide sites for the initiation of long-term hydrolytic degradation by hydrogen ions.

Antioxidants are usually only involved at low level as basic processing stabilizers. Thus some residuals from organic phosphites and sterically hindered phenols may be present. Generally sterically hindered amines are not used, since they would promote aminolysis, a reaction related to hydrolysis.

### 3.2.2 Accelerated testing

Accelerated tests are performed by immersing the PET products in hot water and measuring the reduction in strength, this being the property of most interest. If the environment is to be neutral careful attention has to be paid to the purity of the water and to constancy of temperature. Alkaline contamination of the water must be avoided at all costs. Further Schroeder (1999, 2000) and Schmidt (1994) tested the PET yarns wrapped around cylinders at constant tension to get very low standard deviations for tensile testing. By this procedure the length of the yarns should stay constant i.e. shrinkage is avoided.

Such tests were performed e.g. by Burgoyne and Merii (2007), Schmidt et al (1994), Elias et al (1998, 1999) and Schröder et al (1999, 2000) and Nait-Ali (2009). As hydrolysis is a very slow process the tests have to be continued for more than a year for a statistically significant reduction in strength to be registered at a lowest temperature of around 50°C. Tests can easily be performed above the boiling point of water e.g. Schmidt et al ventured into the region of superheated steam in an autoclave, but under these conditions deviations from the linearity of Arrhenius occur. It is a special problem of PET, that its glass transition temperature in the wet state for the slow internal hydrolysis is in the range around 50 °C and the impact of this transition on hydrolysis is still not fully clear. In several of the tests, therefore, the authors measured the reduction in molecular weight by means of the intrinsic viscosity instead of, or in addition to, measuring the reduction in strength. Separate tests can if necessary be performed to relate molecular weight and/or viscosity to strength.

In the analysis of Burgoyne and Merii (2007) an Arrhenius analysis is applied to the rate at which breaks are created in the polymer chain and this is related to the change in molecular weight. Two yarns were tested, coded PET1 and PET2. The reductions in strength predicted by their calculations for a lifetime of 100 years at 20 °C are given in table 3.1.

In the tests of Schmidt et al (1994) Arrhenius' formula was applied directly to the observed reductions in strength. Predictions derived from their data (including a corrected value for 50 °C) are listed in table 3.1. They noted that the slope of the Arrhenius diagram is slightly greater for the measurements made below 100 °C, leading to a different prediction. As usual, slight differences on a logarithmic scale lead to huge differences in predicted times. Molecular weight was monitored and an analysis based on their measurements but analysed using Burgoyne and Merii's procedure has been added to table 3.1.

Table 3.1 Predictions of strengths of polyester fibres retained after immersion in water.

| | Molecular weight $M_n$ | CEG count meq/g | Method of analysis (see note) | Percentage of strength retained after 100 years at temperature in °C | | | | | Activation energy, kJ/mol | Ref |
|---|---|---|---|---|---|---|---|---|---|---|
| | | | | 20 | 25 | 30 | 35 | 40 | | |
| PET1 | 30700 | 15 | a | 97 | 94 | 89 | 80 | 68 | 112 | Burgoyne, Merii |
| PET2 | 39500 | 30 | a | 94 | 88 | 78 | 64 | 49 | 114 | Burgoyne, Merii |
| 146S | 25300 | 24 | b | 94 | 89 | 79 | 61 | 27 | 98 | Schmidt |
| 146S (<100 °C) | 25300 | 24 | b | 90 | 83 | 69 | 47 | 10 | 85 | Schmidt |
| 146S | 25300 | 24 | a | 98 | 96 | 92 | 85 | 70 | 120 | Schmidt |
| P5 (inferior quality) | 18200 | 47 | b | 69 | 46 | 8 | 0 | 0 | 79 | Elias |
| P6 (PVC coated) | 30200 | 19 | b | 95 | 90 | 81 | 63 | 32 | 100 | Elias |
| P7 | 30600 | 18 | b | 91 | 83 | 69 | 43 | 0 | 94 | Elias |
| Yarn Y1 | | 28 | b | 99 | 97 | 94 | 88 | 76 | 106 | Schröder |
| Yarn Y2 | | 13 | b | 97 | 94 | 81 | 52 | 0 | 108 | Schröder |
| Yarn Y3 | | 26 | b | 99 | 97 | 93 | 86 | 46 | 112 | Schröder |
| Geogrid | | 25 | b | 95 | 90 | 81 | 69 | 44 | 97-109 | Schröder |
| Nonwoven | | 27 | b | 95 | 91 | 82 | 67 | 53 | 100-107 | Schröder |

*Method a: Arrhenius formula applied to the rate of chain breakage;*
*Method b: Arrhenius formula applied to rate of reduction in strength.*

Elias et al (1998) performed similar tests on three yarns, one of which (P5) was deliberately of a lower quality. They analysed the results by applying Arrhenius' formula directly to the strengths. Similar tests like Schmidt were performed by Schröder et al (2000). Predictions based on the authors' calculations are given in table 3.1.

These analyses assume that strength decreases nearly linearly with time at ambient conditions. Since however rupture is probably mainly promoted by end groups on the polymer chains or the related inner hydrogen ion concentration – thus the importance of the carboxyl end group count – the more chains that rupture, the more end groups there are to promote further rupture. This positive feedback, known as "autocatalytic" (Zimmermann 1978) is not significant above about 70% of tensile strength, but could cause more rapid degradation as the strength decreases below this level. Jailloux et al (2008), however, report that the rate of degradation in fact decreases because of an increase in crystallinity. These authors also note an incubation period at the start of immersion during which no reduction in strength is observed. The incubation period lasts 11 days at 95 °C and 35 years at 23 °C, accounting for the lack of observations of any changes on site.

From the results it can be seen that hydrolysis in water for this type of PET-yarns and the qualified other products tested is very slow, so that the retained strength after 100 years at 20 °C for fibres with Mn >25000 and CEG count of <30 meq/g is predicted to be not less than 90% of their initial strength in a saturated chemically neutral soil (100 % r.h.). Above 35 °C the rate of degradation increases rapidly and the predictions become more and more divergent. The assessment yet cannot be done for arbitrary PET products alone on the single basis of molecular weight or CEG count as has been stated above.

### 3.2.3   Index tests and default factors

On the basis of these results, in particular those of Elias et al., ISO TR 20432 sets a requirement that Mn >25000 and CEG count of <30 meq/g for a design life of 100 years. For such materials and in the absence of any other data, it allows a default factor $RF_{CH}$ = 1.2, corresponding to a retained strength of 83%, for a lifetime of 100 years at 25 °C. At 35 °C this is increased to 1.4, corresponding to a retained strength of 71%, but for 25 years only. Arrhenius testing is still recommended for a valid life prediction. The default factor of $RF_{CH}$ = 1.2 corresponds to an activation energy of 107 kJ/mol, assuming a constant rate of loss of strength.

An alternative to this requirement is that the product passes EN 12447, in which it has to have a retained strength of 50% after immersion in water at 95 °C for 28 days (it must be noted that EN 12447 relates to 25 years).

### 3.2.4   Dry and humid soils

Since the degradation of polyesters requires the presence of water, liquid or vaporous, it has to be assumed that no degradation takes place in totally dry soils. However real dry soil possibly exists not even in deserts, the presence of water vapour is enough to start the hydrolysis process. No studies have taken place on alternative, slower forms of degradation, like oxidation, that might occur too in the complete absence of water, perhaps because such environments are rare.

The rate of degradation in humid or partially saturated soils is assumed to be proportional to the relative humidity, or to the degree of saturation. Thus where liquid water is present the local humidity may approach 100% and the degradation should proceed at the same rate as in water. Where the water content of the soil is 80% of the saturated soil it is assumed that the relative humidity is also 80% and that the rate of degradation is similarly 80% of that in water.

Schröder et al (2000) performed some tests at 75 % humidity at 90 °C and showed that the rate of degradation was 60% of that in water at the same temperature. The assumption that the rate of degradation is proportional to the relative humidity is therefore conservative.

### 3.2.5    Coatings and sheaths

The primary purpose of a coating or sheath on polyester geosynthetics is to protect the fine fibres from mechanical damage and weathering. It is suggested that the coating will also restrict outer chemical attack in the same way that paints and other polymer coatings protect metal surfaces from corrosion. The access of water vapour is generally not on the long term excluded. By the coating it will be not possible that low molecular hydrolysis products, which may catalyze hydrolysis since they are composed of compounds with carboxyl endgroups strongly contributing to the carboxyl count , will be eliminated from the product by diffusion or convection thus even an accelerated hydrolysis might occur in the long term view.

Schröder et al (2000) demonstrated that a PVC coating provides no protection from long-term inner hydrolytic attack. Elias et al's (1998) PVC coated yarn P6 performed better than their other yarn P7, although this was not a direct comparison. HDPE coatings are permeable to water vapour in the long term, slowly allowing water under appropriate conditions to accumulate in the narrow space surrounding the fibres. The current view is that internal hydrolysis would be delayed but not prevented in the long term. Impacts on external hydrolysis by large concentrations mineral ions such as calcium might be prevented by coatings with greater success.

Since the effectiveness of the coating cannot be guaranteed in the long term, hydrolysis tests should be performed on uncoated rather than coated fibres as a worst case.

Any increase in predicted lifetime based on a claim that the coating will protect the fibres from chemical degradation must assume that the integrity of the coating is maintained. There must be no cuts or holes or even pores. Assurance of this depends on the backfill used and the quality of the installation process and the quality of the coating. In order to provide this assurance it is essential to perform representative damage tests and systematically record any visible damage in addition to the routine measurements of strength loss. This is not an easy task!

More conservatively thus, it is recommended to base any life prediction on measurements made on the uncoated PET fibres and bars alone.

### 3.2.6    Predictions at higher temperatures

The columns for 30 °C and above for table 3.1 show that the predictions for retained strength at these temperatures diverge. Thus for temperatures up to 35 °C ISO TR 20432 only gives a default factor for 25 years. Any predictions of retained strength, or of RFCH, made for temperatures over 30 °C for durations of over 25 years should be treated with caution.

On the other hand one should question whether such temperatures are really continuous, or whether they represent peak daytime temperatures at the surface in the hottest season. At these times the soil at these temperatures may be relatively dry. It may be possible to reconsider the specification for the environment and avoid an over-pessimistic prediction for the use of polyesters - or any other materials. For the assessment of the effective durations of the elevated temperatures, Arrhenius weight factors have to be included.

## 3.2.7   Predictions for alkaline soils

As described above, in alkaline soils a second form of degradation occurs at the same time as internal hydrolysis. This 'external' hydrolysis results in the erosion of the fibre surface, reducing the strength proportionally to the change of the cross-sectional area, if no local corrosion takes place.

The impact of alkaline hydrolysis is strongly depending on the transport by convection and diffusion of hydroxyl ions to the geosynthetic. This will only happen in wet soil surrounding. Further the impact of carbon dioxide, originating mainly from soil biology, has to be regarded for its neutralizing effects, even if this is not so easy without further local information.

There exists evidence that preferred external hydrolysis takes place around titanium oxide and antimony trioxide ($Sb_2O_3$) particles at the fibre surface. These mineral can usually be present in PET-fibres to control crystallisation.

All four papers that describe accelerated tests on hydrolysis also describe tests in alkaline solutions. Burgoyne and Merii (2007) tested their fibres at 70 °C in $Ca(OH)_2$ (pH estimated as 9 or less) and NaOH at pH 11 for durations of 126 to 250 days. There was no significant difference between the strength of fibres in these solutions and that of fibres immersed in water. Schmidt et al (1994) concluded similarly that at 50 °C and below the rate of reduction of strength in $Ca(OH)_2$ at pH 10 is not different to that in water, i.e. that the internal hydrolysis due to the water dominates over the external hydrolysis due to alkaline attack on the surface. At 70 °C and above, however, the strength diminishes faster in the alkaline solution.

*Schröder et al (1999) performed an extensive series of immersion tests on a woven polyester in neutral (pH 8.1 to 8.3) and slightly alkaline (pH 9.0 to 9.4) soils and in leachates from those soils. The temperature was 60 °C and the durations from 4 to 104 weeks. In the leachates there was an immediate loss in strength of 9%, while after 104 weeks the total loss in strength averaged 19%, an additional reduction of 10%, with a standard deviation of 14%. The immediate loss in strength is dismissed as being due to damage of the specimens by incorporated sharp edged soil particles which damaged the specimen during the tensile test procedure. For the time-dependent loss the authors quote a conservative range of 13 to 24%. There is no significant difference between the neutral and alkaline soils. Meanwhile the CEG count increased from 24 to 38, indicating the presence of internal hydrolysis, while the molecular weight reduced. The exposure at 60 °C for 2 years corresponds to a duration of 185 years at 25°C. For this duration a reduction of 10% would give $RF_{CH} = 1.1$ while the* *standard deviation of 145 would give $R_2$ = 1.16. Reducing this proportionally to 100 years, $RF_{CH}$ = 1.05 and $R_2$ = 1.09.*

*Schröder et al (2000) observed no difference up to pH levels of about 9,5 (hydroxyl ion concentration of 3 × $10^5$ M), beyond which alkaline (external) hydrolysis started to etch the surface while internal hydrolysis degraded the interior of the fibre. They derived an activation energy for alkaline hydrolysis to be 80 kJ/mol for yarn Y1 at 40 to 60 °C, 52 and 46 kJ/mol for yarns Y2 and Y3 at 25 to 60 °C, which is markedly less than the range for internal hydrolysis.*

Vouyovitch van Schoors et al (2009), performing tests in sodium carbonate ("soda") $Na_2CO_3$ at pH 9 and pH 11 at temperatures between 45 °C and 75 °C, concluded that the rate of degradation at pH 11 was no faster than at pH 9. The loss of strength was due primarily to the reduction in molecular weight, while at pH 11 this was accompanied by a reduction in diameter. The modulus decreased by over 10% after two years at only 45 °C and this decrease is expected to occur even at 20 °C after longer durations.

There exists extensive literature on external hydrolysis of PET-fibres. The process is used industrially to etch the surface of PET fibres to get a surface favourable for water vapour transport and eudermic climate in textiles.

These authors all agree that at 20 °C no significant degradation occurs at pH levels of up to 10, even with calcium ions which are more aggressive than sodium (Schröder April 1999).

Elias et al (1998), however, obtained very different results. Tests in $Ca(OH)_2$ were inconclusive due to the formation of deposits, while those in NaOH at pH 10 gave the following predictions for the percentage retained strength after 100 years:

Table 3.2 Predicted retained strengths after immersion in pH 10 for 100 years.

| | Mn | CEG count | Method | Percentage of strength retained after 100 years at temperature in °C | | | | | Activation energy, kJ/mol | Ref |
|---|---|---|---|---|---|---|---|---|---|---|
| | | | | 20 | 25 | 30 | 35 | 40 | | |
| P5 | 18200 | 47 | b | 23 | 0 | 0 | 0 | 0 | 71 | Elias |
| P6 (PVC coated) | 30200 | 19 | b | 83 | 71 | 49 | 14 | 0 | 83 | Elias |
| P7 | 30600 | 18 | b | 65 | 39 | 0 | 0 | 0 | 83 | Elias |

On this basis it is wise to be cautious in specifying polyester for alkaline soils with pH levels markedly above 9, particularly at temperatures greater than 25 °C, although ISO TR 20432 specifies default factors for a pH between 8 and 9. As pointed out in chapter 2.2, the alkalinity of natural soils rarely exceeds 9. pH levels of 10 and above are found only in areas of intense industrial contamination such as mine tailings and in meliorated soils by treatment with lime.

## 3.2.8    Experience from site

Most studies on polyester to date suffer from the difficulty of differentiating at most 1-2% hydrolytic degradation from the effect of installation damage (Troost et al 1994, Elias et al 1998, Cowland et al 1998, Voskamp et al 2001, Naughton and Kempton 2006, Harney and Holtz 2006).

Leflaive (1988) described a comparison on polyester straps taken from a five metre high vertical wall in Poitiers, France, constructed in 1970. The straps had been embedded in the concrete facing elements and anchored in the backfill, where the pH was 8,5. The 40% reduction in strength at the point where the straps entered the concrete facing units was attributed to local high alkalinity (25%), internal hydrolysis (5-10%) and installation damage. Away from the facing elements the reduction in strength was only 2%. The effect of alkaline hydrolysis is clearly strongly depending on the local situation of transport of alkaline liquid extracts originating from the concrete facing to the geotextile, since alkaline hydrolysis will only proceed with significant velocity in the wet state. The effects described by Leflaive can vary strongly e.g. by intensity of rainfall and thus can only be a remark but not a quantitative advice for the designer. Further the important neutralizing effect of carbon dioxid, originating from soil air in an often much higher partial pressure than from ambient air, can be considered only difficultly.

Blume and Alexiew (1998) examined the degradation of a woven polyester after 14 years at the base of an embankment. Taking out the effect of installation damage, it was estimated from changes in Mn and CEG count that 1-2% of the loss in strength could be attributed to chemical degradation. 1-2% over 14 years would amount to a loss of 7-14% over 100 years, assuming a linear rate of loss of strength. Elias (2000) observed a small amount of hydrolytic degradation in polyester geogrids but suggested that this would only become significant and clearly recognizable after 30 years.

Schröder (1998) has reported on excavations of woven polyester geotextile after about 20 years in site, followed by accelerated testing in the water taken from the soil at the same location as the sample. The soil water turned out to be no more aggressive than pure distilled water. The retained strength was approximately 70% after immersion at 90 °C for 84 days and the CEG count increased accordingly. The changes of carboxyl count after excavation was found to be statistically insignificant. The hydrolysis behaviour of the exhumed PET-yarn was identical with a well known PET-yarn studied before. Since molecular mass and carboxyl count were also the same and the tensile behaviour was identical, it was concluded that life expectancy of the exhumed yarn was identical to that of the well known yarn, where excellent longevity had been proved before (Schröder April 1999).

### 3.2.9 Degradation of polyamides

The principal polyamides used in geotextiles are the aliphatic polyamides 6 and 6.6. These have a simple linear chain structure (figure 3.19), the numbers referring to the number of Methylen ($CH_2$)-links between the amide bonds (-NH -CO-).

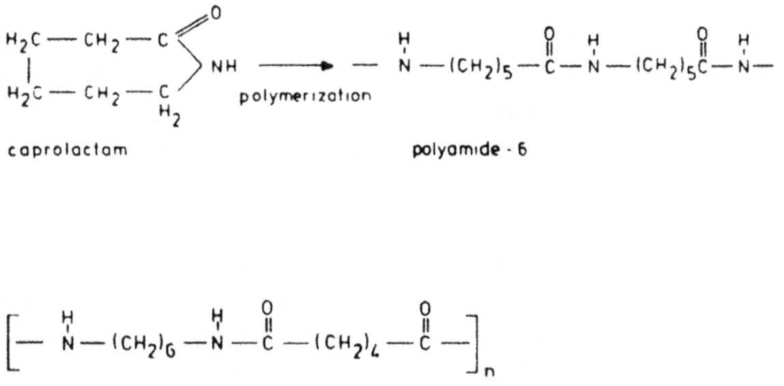

caprolactam                                      polyamide · 6

Fig. 3.19 Synthesis and chemical structure of polyamide 6 and structure of polyamide 6.6 (below), ref Veldhuijzen Van Zanten 1986.

Aromatic polyamides, see figure 3.20, better known as aramids, include benzene rings in the polymer chain which are directly neighboured to the amide bond. The consequence of this situation for durability is that the resistance to oxidation is much improved, since the Methylen-link responsible for oxidation behaviour is missing. But the benzene rings strongly increase sensitivity to weathering. Hydrolysis is still possible and has to be assessed.

Aliphatic polyamides are susceptible to thermal degradation, oxidation, ultraviolet radiation, acid or alkali attack and by hydrolysis through contact with water especially at elevated temperatures. Polyamide geosynthetics are generally markedly more durable to hydrolysis under neutral and alkaline conditions than PET geosynthetics. Under acid conditions they are clearly much more sensitive to hydrolysis and additionally to oxidation than PET. This is because of reasons of fundamental chemical structure and properties of the essential ester and amide bonds. They do not absorb UV light directly, but ubiquitous impurities act as sensitizers like for polyolefins. Oxidation occurs by a chain reaction, very similar to the oxidation of polyolefins but additionally a strong dependence on pH exists. Again in fine fibres oxygen diffusion is retarded by the high degree of crystallinity and orientation and durability is increased correspondingly. Degradation is observed as an increase in yellowness. Sensitivity increases under wet and acid conditions (EN 14030 test method A).

Wet soils have two effects. Most noticeable is the absorption of water – submerged in water of 20 °C approximately 10 % by weight for both polyamide 6 and for polyamide 6,6 while only 4 % at rel. humidity of 65 % at 20 °C (van Zanten 1986). This process is reversible. This absorption of water changes the mechanical properties and dimensions of the polymer by swelling, making it more flexible and facilitates internal hydrolysis. Aliphatic polyamides can be stabilized against oxidation by copper salts, aromatic amines and hindered phenolic antioxidants which all act as heat stabilizers and antioxidants. Hindered phenol antioxidants are the most effective as they inhibit thermoxidative degradation. As with polyesters, hydrolysis stabilizers may be introduced but still it is most important to ensure that the polyamide resin remains dry at the very early stage during processing.

The rate of hydrolysis of polyamides could, in principle, be tested by immersion in hot water or aqueous media of appropriate pH and ionic concentrations in the same manner as for polyesters. Usually the tendency for oxidation may turn out to be even more relevant, especially in acidic aqueous media or wet soil surroundings. Thus hydrolysis testing should be performed under defined oxygen pressures in media of appropriate ionic concentrations (presence of low concentration, about 1mM of ferri ions is recommended, since otherwise the oxidation reaction might prove to be unreproducible because of effects of undefined catalyzing actions of traces of heavy metal ions present) and pH and the procedure should be structured in principle as proposed later for aromatic polyamides. It is also a problem of both aliphatic polyamides, that their glass transition temperature is in the range around of 30 to 60 °C and the impact of this transition on durability related reactions is still not fully clear. CE-marking for 25 year durability requires the index tests for hydrolysis (EN 12447) and oxidation (EN ISO 13438). No results are available to demonstrate whether these methods could be extended to provide a long-term lifetime prediction.

## 3.2.10 Degradation of polyaramids

Polyaramide fibres have been used as reinforcements in composite materials for 40 years. They have been used in geogrids for applications such as the support of railways on piles over old mineworkings, where the modulus of polyester is too low. They are known to be more sensitive to UV light and believed to be more resistant than PA 6 to degradation by oxidation, acids, alkalis and hydrolysis. They are stabilized by adding chlorine and by nitrogen substitution in the polymer chain.

Correspondingly, most studies of the long-term properties of polyaramide fibres have been on their mechanical properties. Recently, Derombise et al (2008, 2009) have reviewed studies of their chemical resistance in humid and alkaline environments over one and a half years. In humid environments degradation is predominantly by hydrolysis and the rate of reaction increases with temperature and applied load. Molecular weight diminishes with the logarithm of time and the associated change in intrinsic viscosity is linearly related to reduction in strength. While other changes were noted, particularly at the fibre surface, they do not appear to be as significant.

n $NH_2$ - ø -COCl → -(NH- ø -CO)$_{n-}$ + $n$ HCl

With ø (benzene ring) =

or:

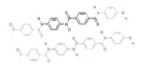

Fig. 3.20 Chemical structure of a typical polyaramide (above synthesis equation, below the structural formula of two sections of a polyaramide chain showing the possibility of relatively strong forces of interchain coupling by the formation of Hydrogen(H)-Oxygen(O) bonds, shortly called hydrogen bridging. Further the tough and stiff structure without these flexible Methylen groups of aliphatic polyamides( PA 6. PA 6.6) is responsible for their stiff tensile behaviour.).

Given an agreed end point (e.g. 50% reduction in strength), the lifetime of polyaramide fibres in humid alkaline environments could be determined as follows:
- Immersion in an alkaline solution at the required pH at elevated temperatures
- Measurement of changes in intrinsic viscosity and mechanical strength
- Derivation of dependence of intrinsic viscosity on time (e.g. on log time)
- Establishment of the relation between intrinsic viscosity and strength
- Derivation of the end point for intrinsic viscosity and hence the predicted lifetimes at the test temperatures
- Establishment of an Arrhenius' relation between lifetime and test temperature and extrapolation the service temperature.

This procedure may be applied in principle to other materials, too.

**Summary**

- The high quality polyester yarns used in geotextiles are generally of similar quality.
- The principal mechanism of degradation in neutral and acid soils is by inner hydrolysis which leads to a slow reduction in strength. All the tests quoted predict at least 90% retained strength for this quality of yarns in normal soil at 100% humidity at 20 °C after 100 years.
- EN 12447 is an index test to assure durability for a 25 year lifetime at temperatures up to 25 °C in soils of pH between 4 and 9.
- ISO TR 20432 specifies the same index test for a 100 year lifetime, with the alternative of $M_n$ >25000 and a carboxyl end group (CEG) count of <30 meq/g. In this case the default reduction factor for the same soil condition is 1.2. Accelerated tests consisting of immersion in water at high temperatures are however recommended for life prediction.
- Degradation occurs more rapidly at higher temperatures, particularly at 35 °C and above, and the predictions of retained strength diverge according to the source and the method of prediction.
- Degradation will be slower in drier soils and should be retarded by any sheath or coating, although this cannot be relied on as it may be damaged during installation.
- In alkaline soils additional 'external' hydrolysis takes place, and predictions of its effect vary widely. In soils with higher pH values polyester based reinforcements should be used with caution for long-term applications until more reliable information is available.
- Polyamides and less the closely related polyaramides degrade by oxidation as well as by hydrolysis. No long-term data exist, although methods are available for lifetime prediction.

# References

*Chapter 3.2*

- Ludewig H., Chemie und Technologie von Polyesterfasern, Akademie Verlag Berlin, 1975, p. 131.
- Ravens D.A.S., Ward I.M., Trans.Farad. Soc., 57, pp 150.
- Zimmermann H., Oct. 1978, Soc.of Plast. And Rubber Eng., 2nd Int. Symp.on Degr. and Stab. of Polymers, Dubrovnik, paper 02.
- Akers S A S, Studinka J B, Meier P, Dobb M G, Johnson D J, Hikasa J.; Long term durability of PVA reinforcing fibres in a cement matrix; The International; Journal of Cement composites and Lightweight Concrete, Vol. 11, No. 2, 1989, pp79-90.
- Blume K-H, Alexiew D.; Long-term experience with reinforced embankments on soft subsoil: mechanical behaviour and durability. Sixth International Conference on Geosynthetics, Atlanta GA, USA, 1998, pp 663-668.
- Burgoyne C, Merii A.; On the hydrolytic stability of polyester yarns; Journal of Materials Science Vol. 42, pp2867-2878, 2007; Originally published in 1993 as a Cambridge University Engineering Department Report.
- Cowland J W, Yeo K C, Greenwood J H.; Durability of polyester and polypropylene geotextiles buried in a tropical environment for 14 years; Sixth International Conference on Geosynthetics, Atlanta GA, USA, 1998, pp 669-674.
- Derombise D, Van Schoors L, Davies P. Durability of aramid geotextiles in an alkaline environment, 2008. EuroGeo4, Edinburgh, UK, paper 287, pp 1-7.
- Derombise G, Vouyouvitch van Schoors L, Duragrin D, Barberis N, Davies P.; Vieillissement des fibres aramides utilisées dans les géotextiles pour le renforcement des sols alcalins; Rencontres géosynthetiques, Nantes, 2009, pp 149-156.
- Elias V, Salman A, Goulias D. The effect of pH, resin properties and manufacturing process on laboratory degradation of polyester geotextiles, 1998. Geosynthetics International, Vol. 5, 459-490.
- Elias V, Salman I, Juran I, Pearce E, Lu S, "Testing protocols for oxidation and hydrolysis of Geosynthetics", US Federal Highways Administration Report FHWA-RD-97-144, published 1999.
- Harney M D, and Holtz R D. Mechanical properties of reinforcement, 30 years after installation, Eighth International Conference on Geosynthetics, Yokohama, Japan, 2006. Millpress, Rotterdam, pp 1041-1044.
- Hirai T, Hikasa J, Kishitani K. ;Durability study of fiber cement reinforced with PVA fiber.
- JCA Proceedings of Cement and Concrete, No. 47, 1993, pp 618-623 and 652-657.
- Hsuan Y.G., Schroeder H.F., Rowe. K., Müller W., Greenwood J., Cazuffi D., Koerner R.M.; Long-Term Performance and Life Time Prediction of geosynthetics, EuroGeo4 Keynote Lecture, Edinburgh, 2008.
- Nishiyama M, Yamamoto R, Hoshiro H. Long-term durability of Kuralon (PVA fiber) under alkaline condition. International Inorganic-Bonded Fiber Composites Conference (IIBCC), Sao Paolo, Brazil, 2006. pp120-133.

- Jailloux J-M, Nait-Ali K L, Freitag N. Exhaustive long-term study on hydrolysis of high-tenacity polyester – 10 year results, 2008. EuroGeo4, Edinburgh, UK, paper 212, pp 1-6.
- Leflaive, E., (1988), "Durability of Geotextiles: the French Experience", Geotextiles and Geomembranes, Vol. 7, pp 23-31.
- Newman, E., Stark, T.D., Rohe, F.P., (2001), "PVC Aquaculture Liners Stand the Test of Time", Geotechnical Fabrics Report, IFAI, Roseville, MN, USA, September 2001, pp 16-19.
- Naughton P J, Kempton G T. Lifetime assessment of polyester based geosynthetics. Eighth International Conference on Geosynthetics, Yokohama, Japan, 2006. Millpress, Rotterdam, pp 1577-1580.
- Nait- Ali K L, Thomas R W. Anderson R L, Freitag N; Hydrolysis testing of high tenacity Poly(ethylene terephthalate) – Results of 15 years of exposure; Geosynthetics 2009, Februar 2009, Salt Lake City,l Utah.
- Schmidt H M, Te Pas F W T, Risseeuw P, Voskamp W.; The hydrolytic stability of PET yarns under medium alkaline conditions; 5th International Conference on Geotextiles, Geomembranes and Related Products, Singapore, 1994. International Geotextile Society, 1994, pp 1153-1158.
- Schröder H F, Lorenz E, Bahr H, Seifert I, Neyen V, Benemann S, Strauss B, Volland G, Lange B. - Untersuchungen zum Langzeit-Hydrolyse-Verhalten von Polyestergeweben.; 6. Informations- und Vortragstagung über "Kunststoffe in der Geotechnik", Munich, March 1999, ed. Floss R. Deutsche Gesellschaft für Geotechnik e.V., Essen, Germany. pp 291-305.
- Schröder H F, Bahr H, Kneip G, Lorenz E, Schmücking I, Seifert S, "Long-term resistance of PET geofibers to inner and outer hydrolysis", China International Nonwovens/ Techtextiles Conference, Beijing, 2000, pp 165-183.
- Schröder H.F. Abschlußbericht zum Forschungsvorhaben des Deutschen Institutes für Bautechnik (DIBt) Az: IV 1-5-666/92 Nr.11.40, 23. April 1999.
- Troost G H , den Hoedt G, Risseeuw P, Voskamp W, Schmidt H M.; Durability of a 13 year old embankment reinforced with polyester woven fabric, 5th International Conference on Geotextiles, Geomembranes and Related Products, Singapore, IGS, 1994, pp 1185-1190.
- Voskamp,W., Van Vliet,F. and Retzlaff,J. Residual strength of PET after more than 12 years creep loading, Proc. of the International Symposium on Earth Reinforcement (Ochiai et al., eds), Balkema, 2001, Vol.1, pp.165-170.
- Vouyovitch van Schoors L, Lavaud S, Duragrin D, Barberis N., Durabilité des géotextiles polyester en milieu modérément alcalin, Rencontres Géosynthetiques, Nantes, France, 2009, pp 201-207.
- Van Zanthen, R. Veldhuizen, Geotextiles and Geomembranes in Civil Engineering, A.A. Balkema/Rotterdam/Boston/ 1986, p. 105.

## 3.3     Weathering

J.H. Greenwood, H.F. Schoeder, P. Trubiroha

### 3.3.1     Weather

Solar radiation consists of a range of wavelengths extending from 300 nm to about 2500 nm. The section from 400 nm (violet) to 800 nm (red) is visible to the naked eye and is called "light". The radiation can also be regarded as individual photons whose energies are highest at the shorter wavelengths and lowest for the longer ones. Particularly significant for photo-degradation are those with the highest energies, corresponding to the 'ultra-violet' or UV radiation which forms part of the solar spectrum but which is invisible to the naked eye. UV radiation is subdivided into UV-B, with wavelengths below 320 nm, and UV-A, with wavelengths between 320 and 400 nm.

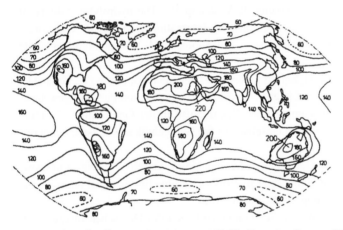

Fig. 3.21 Map of radiant exposure per year, Veldhuijzen van Zanten, 1986, the values shown have to be multiplied by 42 to get MJ/m² (note: 100 kcal/cm² = 4.2 GJ/m² ).

As would be expected, solar radiation is highest at tropical latitudes. At temperate latitudes it is less intense and varies more with the seasons. In the polar regions not only do the long days of summer and the long nights of winter lead to a more extreme seasonal variation, but also depletion of the ozone layer has led to a higher proportion of UV radiation. UV radiation is also greater at high altitudes.

The countries of Southern Europe experience an annual total radiant exposure of 4 GJ/m² to 6 GJ/m², thus a mean value of 5 GJ/m² is assumed. The UV radiant exposure (radiation with wavelength < 400 nm) is 6% or 300 MJ/m² per year.

For Central Europe an annual total radiant exposure of 3,5 GJ/m² to 4,2 GJ/m² and for Northern Europe 2,5 GJ/m² to 3,5 GJ/m² with the UV radiant exposure of 6 % for both can be assumed.

As we know in Central Europe, weather is not just solar radiation. Clouds allow radiation to pass through, and the proportion of UV in this diffuse radiation is somewhat higher than in the direct solar radiation itself. Rain follows sun follows rain, and these cycles of wet and dry can be as important to a geosynthetic as the volume of rain itself. Even without falling rain the humidity varies widely, for example condensing as dew on colder surfaces under a clear sky at night. In spite of the finest modern computing and the sweet utterings of weather girls, the weather will always surprise us. Its variability is reflected by the number of words in the English language that are used to describe it.

## 3.3.2   Effect of weathering on geosynthetics

The high energy of the individual photons of UV radiation is just sufficient to produce changes to the polymer itself. These changes are complex and vary from one polymer to another. Some are summarised in the following table.

Note that the values are activation spectra determined with the filter technique and are specific both for the property change and for a distinct radiation source and its filter system. The wavelength (ranges) in the table show the area in which the product of spectral irradiance and spectral sensitivity is the highest (for this property change). Their use requires information about the lamp and its filter system (Trubiroha et al 2005). Generally the spectral sensitivity (action spectra) of polymer degradation increases with decreasing wavelengths which means with increasing photon energy.

Table 3.3 Effect of weathering on geosynthetics

| Polymer | Most sensitive wavelength | Type of degradation |
| --- | --- | --- |
| PE | 330 to 360 nm | Photo-oxidation is initiated by the generation of free radicals, which requires less energy than breakage of a bond. These radicals can then start a chain oxidation reaction leading to breakage of the polymer backbone. |
| PP | 355 to 360 nm | |
| PVC | 320 nm | Bond breakages initiate the chemical degradation of PVC by elimination of HCl and the formation of polyenes. Some oxidation takes place at the surface. The UV can also affect plasticizers. |
| PET | 325 nm | Bond breakage in the main chain (chain scission) leads to reduction in strength and the generation of carboxyl groups. |
| PA | The spectral sensitivity of PA 6 depends on the degree of crystallinity (Trubiroha et al 2005). A higher degree of crystallinity makes PA 6 more stable. | |

Much of the literature on the weathering of geosynthetics relates to geomembranes, since these are more likely to be exposed to solar radiation for very long durations, for examples in reservoirs or in liquid storage. This experience is relevant to polymers made from similar materials and will therefore be referred to in the text that follows.

### 3.3.3   Changes to polymers due to UV radiation

In most cases the UV radiation will produce changes which, directly or indirectly, lead to breakage of the main chain of the polymer. Some reactions are very sensitive to the wavelength of the incident radiation. This in turn leads to a reduction in strength, reducing the effectiveness of a reinforcement or allowing a geomembrane or filter to tear. Other results, such as changes in colour, often are less important for geosynthetics than they are for furnishings or paints. However, the discoloration of light-coloured geotextiles to a yellowish colour often indicates the destruction of phenolic antioxidants which may be relevant for the life expectancy of the product.

Just as important is the effect of temperature. Visible and infrared radiation of all wavelengths heats the geosynthetic on to which it falls, and much of the degradation caused by radiation is due simply to the rise in temperature of the material. Remember that an increase of 1 °C can lead to an increase of 8% in the rate of change of a property, which means that an increase of 10 °C can, according to van't Hoff's rule (see section 2.3.9.6) may double it. The local surface temperature will be equal to the air temperature added to the effect of radiation. At the Arctic Circle in July the air temperature averages 15 °C while in Dubai it is 41 °C. The additional temperature due to solar radiation is roughly 20 °C at the Arctic Circle and 30 °C in Dubai, making a total difference of 36°C in local temperatures. This means that oxidation could proceed about 12 times faster in Dubai. Grubb et al (2000), reporting on exposure of various polypropylene geotextiles in the high Andes, noted that high temperature rather than radiant exposure was the parameter more closely associated with degradation. They also noted that the lighter materials were more susceptible to degradation, an observation reflected in the stipulation that the results of weathering tests on a light product may be used for a heavier product in the same range, but not vice versa. Conversely, the results of many years of exposure of polymers to UV radiation in the high Alps proved as worthless due to the low local temperatures as they were enjoyable to the staff sent to monitor the tests.

Humidity and local moisture have a particular effect on polyamides, whose mechanical properties are well known to be sensitive to moisture content. An increase in relative humidity from 10 % to 90 % has an effect on PA6 similar to an increase of temperature by 10 °C. Acidic precipitation can also affect polyamides. Samples of a polyamide geosynthetic exposed under what turned out to be similar conditions during a bright summer in Berlin and a poor one in the south of France showed a reduction in strength in Berlin which could be attributed to acid pollution (Greenwood et al 1996). Weathering of Geosynthetics made from PET is relatively insensitive to temperature and humidity, however, while PP is strongly influenced by temperature but not generally by humidity. Additives may influence this behaviour. Local moisture also provides an opportunity for the leaching and hydrolysis of reaction products, radicals and additives.

## 3.3.4   Duration of exposure

In many applications geotextiles are only exposed to radiation during installation and, if unprotected, storage. For functions such as filtration, separation, drainage and the lining of landfills the geosynthetic has of necessity to be covered if it is to perform its function, and the same is true of most reinforcement and erosion control applications. Even reinforcements left exposed at the front of a soil embankment are soon shielded by vegetation. At the edge of reservoirs and canals, to take two examples, a protective geotextile would be exposed to solar radiation throughout its lifetime.

Nevertheless, these few weeks of exposure to solar radiation can lead to significant degradation in a polymer that is insufficiently stabilised. The exposure of the wrapping of the Reichstag in Berlin had to be limited to 14 days, since natural weathering tests the year before had shown that the aluminium coating protecting the woven PP- split-film yarn was so thin that the textile could fail after that time. As a precaution, laboratory tests were performed in parallel which simulated the weathering behaviour under nearly identical conditions in order to ensure that no preliminary failure would occur, (figure 3.22).

Fig. 3.22 Reichstag in Berlin, wrapped in aluminium coated PP split-film yarn woven by Christo (ref Donald Eubank).

The exposure can also lead to changes which may not have an immediately noticeable effect but which will reduce the lifetime of the geosynthetic in the long term (see 3.3.7).

## 3.3.5   Additives

In view of the rapid degradation that can occur in an unstabilized polymer, most geosynthetics contain additives to counter the effect of UV radiation. The most familiar is carbon black which absorbs the UV and restricts its access to the polymer. More finely divided carbon, diameter typically (22-25) nm, is more effective. Thus many geotextiles are black. Some include white pigment such as titanium dioxide (rutile) instead: this absorbs UV radiation and reflects more of the light, while high humidity or moisture can cause photocatalytic degradation. Carbon black and white pigments should not lose their effectiveness with time or be leached out of the polymer.

Other additives such as hindered amine light stabilizers (HALS), which do not absorb in the range 300-400 nm, but deactivate the reactive radicals which would otherwise have accelerated oxidation. In long-term use some of these additives can be consumed or, in the presence of liquid, be leached from the surface. Additives introduced for other purposes, such as long-term stabilisation, can themselves suffer from photo-oxidation, as described in 3.3.8. Additives generally are described more fully in section 3.1.4.

Fig. 3.23 Reservoir lining exposed to full solar radiation (courtesy GSE Lining Technology GmbH).

Fig. 3.24 Geosynthetic mat, filled with grout (ref.: A.S.P. Enterprises Inc).

Most geosynthetics contain sufficient stabilization to survive exposure during storage and installation without significant loss of strength or loss of long term stabilizers. This can be tested using OIT or oxidation resistance methods. Leaching of additives and of the products of photo-oxidation is addressed by including a water spray in the standard test methods.

### 3.3.6 Natural weathering

Outdoor natural weathering is used widely to test polymers and paints; it is easy to set up, cheap to run and represents reality. The principal disadvantage is that, as mentioned, one period of weather is never the same as the next, however well monitored the conditions may be. While it may be possible to plan for the variation between winter and summer in temperate climates or wet and dry seasons elsewhere, our experience of summer holidays tells us that in one year a particular month can be wet and cold and in the following year gloriously sunny. In Hamburg (not a summer holiday location), where the radiant exposure by global solar radiation incident in June averages 450 MJ/m², the values recorded over the last 40 years extend from a minimum of 80 to a maximum of 850 MJ/m². UV radiation makes up about 6 % of this radiant exposure. Consequently, outdoor exposure under apparently similar conditions can lead to a wide spread of results.

The other problem with outdoor exposure tests is that it takes place in real time and is therefore slow, apart from special sites where the solar radiation is concentrated up to eight times by specially coated mirrors. Nevertheless, standards ISO 877 and ASTM D 5970 exist for outdoor exposure tests, during which it is essential to record the total solar radiant exposure and UV radiant exposure.

There have been many results indicating the degradation of geosynthetics exposed to weather. Some degrade within months, while well stabilized and correctly installed geosynthetics have survived for many years. Degradation takes place even under water, but more slowly. Where the surface can be alternatively wet and dry, however, such as in the intertidal region, degradation can be accelerated. All will depend on numerous factors like the thickness of the geotextile, the quality of the stabilization, the intensity and duration of the radiation, the surface temperature, the leaching of additives or the products of reaction. Transfer of results from one location to another can be based, crudely, on the radiant exposure and mean temperature, but a safety factor of at least two should be allowed in order to cover all the possible causes of variation.

### 3.3.7 Accelerated weathering

While natural weathering is slow and the conditions variable, accelerated testing has the advantage that it is faster and, just as important, the conditions are controlled. The geosynthetic is exposed to high intensity UV radiation, which increases the number of photons but additionally the temperature and relative humidity of the sample is controlled, and the duration of exposure is extended from 12 to approaching 24 hours a day. Radiant exposure is used as the basis for comparison with service conditions. Since dark periods and the presence of water play a part in the migration and leaching both of additives and the products of reaction, exposure to radiation is interrupted by short controlled periods of water spray either when dark or light.

The limit to the extent of acceleration is set by the maximum temperature at which degradation of the material can be regarded as representative of the conditions in service. The intensity of the radiation (irradiance level) can only be so high as not to allow the specimen to get unrealistically hot, under the essential rule that the mechanisms operating in the accelerated test must be representative of those occurring in practice. The ratio between the irradiation conditions in EN 12224 and one year's natural weathering ranges from 4 in southern Europe to 8 in northern Europe. This limitation has proved very restrictive, indicating that geosynthetics should be exposed to artificial weathering for up to 10 years if they are to simulate a lifetime of 40 years in southern Europe.

There are two principal types of accelerated weathering equipment, known also (particularly in North America) as weatherometers:

- Xenon arc equipment (ISO 4892-2, ASTM D 4355) has a spectrum that simulates natural daylight, including the UV region. The irradiance is regulated to provide a controlled surface temperature, in so far as this can be applied to geotextiles, and the relative humidity is also controlled. Surface temperatures are clearly higher than the chamber temperature, mean temperature differences are clearly higher than in practice and therefore the photo degradation of dark coloured materials is more accelerated than the photo degradation of light coloured materials.
- Fluorescent UV equipment (ISO 4892-3, ASTM D 7238) uses fluorescent lamps to simulate purely the ultraviolet region of natural daylight, the wavelengths ranging from 300 to 400 nm with a peak at 340 nm. By omitting the visible and infrared radiation the specimen remains cooler, such that the surface temperature is equal to the testing chamber temperature, i.e. is independent of the colour and thickness. This is a disadvantage for the testing of dark coloured materials which show elevated surface temperatures in comparison to light coloured materials in practice. Due to the construction of the equipment the area that can be exposed in the fluorescent UV test is much greater than in rotating rack xenon arc machines. The fluorescent UV test is therefore preferred for geosynthetics. North American and European standards exist for this method and are compared in the following table (Hsuan et al 2008):

Table 3.4

| Standard | Irradiance | Dry cycle | | Wet cycle | | |
|---|---|---|---|---|---|---|
| | | Temperature (°C) | Duration (h) | Temperature (°C) | Duration (h) | Application of water |
| EN 12224 (geotextiles) | 40 W/m$^2$ (total 50 MJ/m$^2$) from 300 to 400 nm | 50 ± 3 | 5 | 25 ± 3 | 1 | Water spray |
| ASTM D 7238 (geo-membranes) | 0.78 W/ m$^2$.nm at 340 nm | 75 ± 3 | 20 | 60 ± 3 | 4 | Condensation |

Fig. 3.25 Xenon arc equipment (courtesy Materials Technology Ltd.).

Fig. 3.26 Fluorescent bulb equipment courtesy Q-Lab).

Fig. 3.27 Fluorescent UV lamps device (courtesy Dr. V. Wachtendorf, BAM 6.6) (On the left view of the opened device the inner space with some exposed specimens and in the door inside the fluorescent lamps can be seen, on the right the enlarged operator controls is shown, which is positioned on the other side wall of the device.)

In both methods the irradiating conditions and the temperature of the specimen are measured and recorded. Tests according to EN 12224 should take about two weeks to complete.

### 3.3.8    *Requirements on geosynthetics exposed only during installation*

The US federal transportation requirements (AASHTO M288) require that geotextiles have 70 % retained strength after exposure for 500 h to ASTM D7238. Lack of reproducibility between laboratories suggests that this requirement may be too stringent for nonwoven materials (Hsuan 2008, quoting Searle and Sandri 2001). Since practically all geotextiles and most geomembranes will be covered after installation, the procedure adopted by European standards is to limit the time that the geotextiles may be exposed on site according to the results of the accelerated weathering test EN 12224, which corresponds to approximately one summer month's exposure to solar radiation in southern Europe. In reality most commercially available geosynthetics have sufficient protection against the ultraviolet to survive a month's Mediterranean sunshine without significant physical damage. The maximum exposure times are set out in the following table.

Table 3.5

| Retained strength after testing to EN 12224 | Maximum exposure duration (uncovered) during installation |
| --- | --- |
| >80 % | 1 month |
| 60 % - 80 % | 2 weeks |
| <60 % | 1 day |
| Untested material | 1 day |

Thus for geotextiles there is no specific life prediction, solely a limit on the time the material is permitted to remain exposed on site. The one month duration in the second row may be increased to a maximum of four months for climates and seasons less balmy than the Mediterranean in summertime to which it refers. RFW, the reduction factor for weathering effects, may be rounded down to 1.00 if the measured ratio is less than 1.05 or is not statistically significant.

Note however that long term stabilisers such as phenolic antioxidants, which are needed to provide resistance to oxidation in the long term, may be affected by the UV radiation if they are not sufficiently protected (Zweifel 1998, Gugumus 1995). The only indication of this is a yellowish discoloration for light-coloured geotextiles. Degradation of the stabiliser, which will not be detected by the EN 12224 test, can be identified by OIT testing of weathered material.

If appropriate information is not available for one product within a range then it is acceptable to use the results for a lighter grade, since this is more likely to be sensitive to weathering. Values from heavier grades should not be used. Care should be taken in transferring data between coated grades where the effect of weathering is more likely to be related to the thickness of the coating than to the strength of the fibres inside it.

## 3.3.9  Lifetime prediction for geosynthetics permanently exposed

### 3.3.9.1  Extrapolation of measurements on samples taken on site

This leaves the question of how to assess geosynthetics, particularly geomembranes, which are to be exposed to solar radiation for their entire service lifetime. The simplest method is to monitor the material in situ, taking samples where possible without compromising the application. This has the advantage that the material and conditions of exposure are, subject to variations in the weather, those for which the prediction is to be made. The problems are that this process is slow, can only be performed once the geomembrane has been installed, requires definition of an endpoint (e.g. 10% retention or total exhaustion of the antioxidant stabilizer), and that the nonlinear rate of degradation leaves it unclear how fast the observed changes will proceed over the future period of prediction.

Several authors show graphs of the change in mechanical properties or of OIT with time to which a linear or nonlinear curve can be fitted. Rowe and Rimal (2008) show that the OIT of a HDPE geomembrane exposed in Canada has reduced from 130 to 50 over 9 years and predict a lifetime of 46 years to complete exhaustion of the antioxidants. Hsieh et al (2006) illustrate a similar nonlinear degradation. Heibaum (2006) predicts a lifetime of 20 to 30 years for an exposed HDPE geomembrane provided that it is properly stabilized and correctly installed. More problems occur in smaller projects without the same quality of design and installation.

### 3.3.9.2   Certification of a new material based on experience

The same provisos and uncertainties exist in the use of data from experience in the certification of a new material for a future application. As a minimum, evidence should be provided giving full details of the reference project, the date of installation and its location and use, together with proof that the product for which a prediction is to be made is identical with the existing reference product in all properties that relate to durability. Samples for test purposes should be taken from "worst case" locations, for example south facing slopes at the water's edge which can be either covered or exposed as the water rises and falls. The samples should be tested for tensile strength and elongation and, where appropriate as in PP and PE, subjected to OIT. The rates of change can then be related to the prediction in hand including the design life. The problems associated with the use of data on similar products are discussed in 3.8.3.

### 3.3.9.3   Accelerated testing

Providing accelerated tests to predict the performance of a geosynthetic has proved to be both difficult and controversial. The current situation is that tests to ensure a reasonable prediction of lifetime may themselves last longer than five years, far too long for the average project to wait. Tests such as EN 12224 can be continued for longer, using as the basis for comparison radiant exposure, that is the total amount of UV radiation experienced, or the time integral of the spectral irradiance. As mentioned, the 50 MJ/m$^2$ specified in the index test EN 12224 is believed to simulate one average summer month's exposure in Southern Europe. This is about one-sixth of the annual UV radiant exposure. Greenwood et al (1996), describing the experimental work on which EN 12224 was based, showed that when based on comparable radiant exposures both the xenon and fluorescent UV methods of accelerated weathering provided a satisfactory prediction of natural weathering, although the prediction tended to be conservative for white geotextiles and non-conservative for black ones. Since the test irradiance is about 40 W/m$^2$ and should be interrupted for one hour in six, the standard test duration is approximately 430 h or 18 days. To effectively simulate average European conditions on site for a period of one year the required test period should be six times this or 2580 h, 3.5 months of testing., this would be reduced to about 2000 h or 2.7 months if the radiation source is not switched off during the spray cycle.

Comparison between accelerated and natural weathering based on radiant exposure has been shown to be broadly correct, although the error in individual cases can exceed 50 %. Temperature, altitude, humidity and the equipment used in real time tests have a significant effect on the correlation.

Comparisons quoted by Hsuan et al (2008) suggest that close control of temperature is necessary. Comer et al (1998) noted cracking in a fPP geomembrane after 26 years of service, and retrospectively showed that this would have been correctly predicted by a fluorescent UV test. Heindl et al (2008) provided a detailed comparison of artificial and natural weathering of a PP geotextile with different levels of stabilization for exposures of up to 180 MJ/m². The comparison between natural weathering and artificial weathering was generally good when compared on the basis of radiant exposure, the comparison being better with the more severe Mediterranean exposures than with those in Northern Europe. If the content of recycled material in geomembranes increases the prediction of long-term resistance to weathering it will become even more complicated. Maxwell (2008) in a study of exposed and unexposed HDPE taken from wheelie-bins in the UK showed that, although the samples had similar properties directly after reprocessing, the samples containing exposed polymer degraded more rapidly, presumably due to the presence of photo-degraded products. This raises questions of how to test the resistance to weathering of geosynthetics containing recycled material.

**Summary**

*When exposed to UV radiation polymers can degrade by photo-oxidation, especially at the surface. This leads ultimately to a reduction in strength.*

Heating caused by the whole spectrum of solar radiation accelerates photo-oxidation as well as other forms of degradation present.

Most geosynthetics are exposed to light solar radiation only during storage on site and installation.

Most geosynthetics contain sufficient stabilization to survive this exposure without significant loss of strength or loss of long term stabilizers. This can be tested using OIT or oxidation resistance methods.

Geosynthetics which are to be covered in use should be tested according to EN 12224 or ASTM D 7238. The result indicates the maximum time they may be exposed to solar radiation during installation. There is no specific life prediction.

• The lifetime of geosynthetics exposed permanently, such as in reservoirs or canals, is more difficult to predict. Methods include:Given sufficient time, samples can be taken on site and changes in key parameters such as elongation at break or OIT can be extrapolated into the future.

• Comparison with experience with a similar material, as described in 3.3.8.2.

Artificial weathering can be performed using xenon arc or fluorescent UV testing equipment according to standard procedures, such as the method of EN 12224 extended to longer durations, and has so far proved a satisfactory simulation of practice, provided that an allowance is made for variability. From examination of existing results, a safety factor of two is recommended. Artificial weathering is limited by the maximum acceptable temperature of the specimen, restricting the rate of acceleration to a factor of 3-4. Tests lasting years are therefore needed to predict commercial service lives lasting decades.

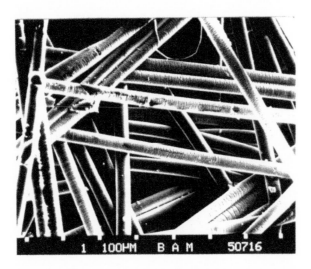

Fig. 3.28 Surface damage of the PA/ PP (sheath/core) fibres of a nonwoven geotextile after about 3 summer months of natural weathering in Berlin -(damage was caused by UV radiation and acid rain).

# References

*Chapter 3.3*

- Comer A I, Hsuan Y G, Konrath L. Performance of flexible polypropylene geomembranes in covered and exposed environments. Sixth International Conference on Geosynthetics, Atlanta GA, USA, 1998, pp 359-362.
- Greenwood J H, Trubiroha P, Schröder H F, Franke P, Hufenus R., Durability standards for geosynthetics: the tests for weathering and biological resistance, Geosynthetics: Applications, Design and Construction; ed. De Groot MB , den Hoedt G, Termaat R. Balkema, Rotterdam, 1996, pp 637-642.
- Grubb D G, Diesing W E III, Cheng S C J, Sabanas R M, Comparison of geotextile durability to outdoor exposure conditions in the Peruvian Andes and the Southeastern USA, Geosynthetics International Vol 7, 2000, pp 23-45.
- Gugumus, F., in: Current Trends in Polymer Photochemistry, Eds. Allen, N.S. Edge, M., Bellobono, I.R. and Selli, E., Ellis Horwood, N,Y.( 1995) 225.
- Heibaum M, Fourie A, Girard H, Karunaratne G P, Lafleur J, Palmeira E M. Hydraulic applications of geosynthetics. Eighth International Conference on Geosynthetics, Yokohama, Japan, 2006. Millpress, Rotterdam, pp 79-120.
- Heindl M, Zanzinger H, Zahn A, Schönlein A Study of artificial and outdoor weathering of stabilized polypropylene geotextiles, Proceedings of the 4th European Geosynthetics Conference, Edinburgh, UK, 2008, Paper 286, pp 1-8.
- Hsuan Y G, Schroeder H F, Rowe K, Müller W, Greenwood J H, Cazzuffi D, Koerner R M. Long-term performance and lifetime prediction of geosynthetics. EuroGeo4, Edinburgh, UK, pp 327-366.
- Maxwell A S Weathering of recycled photo-degraded polyethylene Polymer Engineering and Science Vol. 48, pp381-385, 2008.
- Rowe K, Rimal 2008, Hsieh C, Wang J-B and Chiu Y-F, Weathering properties of geotextiles in ocean environments, Geosynthetics International, Vol. 13 (2006) pp 210-217.
- Searle N, Sandri D, 2001, Weathering stability testing of geotextiles, Geotechnical fabrics report, October/November 2001, pp 12-17.
- Trubiroha, P., Geburtig, A., Wachtendorf, V. Determination of the Spectral Response of Polymers.
- Service Life Prediction - Challenging the Status Quo, Ed: Ed.: Martin, J.W., Ryntz, R.A., Dickie, R.A., 2005, ISBN 0-934010-60-9; p. 241 - 252.
- Zweifel H. Stabilisation of Polymeric Materials, Springer- Verlag Berlin Heidelberg, 1998, p 65.

## 3.4     Tensile creep

J.H. Greenwood

### 3.4.1     Introduction to creep

One function of a geosynthetic soil reinforcement is to support a soil structure by carrying load − load that the soil alone would be unable to carry. There are many others, such as supporting embankments over soft soil or over piles. If the reinforcement breaks the structure will collapse − possibly with catastrophic consequences. Its ultimate tensile strength (UTS) is measured by rapid loading, but the reinforcement may also break following a long duration at a lesser load. This is known as **creep-rupture**, or, less commonly, stress-rupture or static fatigue. The higher the load, the shorter the lifetime. Design to this failure point is known by civil engineers as ultimate limit state design, and in view of its serious consequences much work has gone into predicting the highest load that could be applied continuously to the geosynthetic over the course of its design life without leading to rupture.

But polymer geosynthetics do not just break under sufficient load. When load is applied to a metal or ceramic at a normal temperature the material will extend a small amount and then remain at this length until the load is removed and the material returns to its original length. This is referred to as **elastic behaviour**. Geosynthetics and other plastics, however, extend under load and then continue to extend for as long as that load is maintained. If the load is removed part of the extension will be regained immediately, following which the material will contract, if not necessarily regaining its original length. This is known as **visco-elastic behaviour** or, more commonly, **creep**. Creep is dependent not only on the polymer type and grade, load and temperature, but also on the structure of the geosynthetic and on its method of manufacture.

Creep of the reinforcements could cause a soil structure to slip or sag without total collapse. A limit on the acceptable extension is called a serviceability limit state. The creep-strain data are used to analyse the total strain which takes place during the design life of the structure. The total strain consists of elastic deformation during loading and creep elongation during the design life of the structure.

Typical reinforced embankments are built with slopes steeper than would be possible without reinforcement, in which case limited extension of the geosynthetic after construction is accepted. In these structures the load on the geosynthetic is regarded as remaining constant. In embankments built over soft ground, with reinforcement at the base, the load will increase over time, at least until the soil consolidates, but again may be taken conservatively as constant. In some instances engineers have tried to prestress a geosynthetic (like a steel reinforcement) supporting a wall, bridge abutment or other structure. In this case, therefore, there will be no further extension in the material after the prestress has been applied. Rather than the geosynthetic extending, the stress in the geosynthetic will decrease or relax with time. This is known as **stress relaxation**. Geosynthetic reinforcement layers in reinforced soil walls under operational (serviceability) conditions are likely to undergo both creep and stress relaxation simultaneously, comprising some extension and some reduction in stress (Allen and Bathurst 2002, Kongkitkul et al 2010).

Creep also occurs in drainage geosynthetics, but in compression perpendicular to the material. The pressure of the soil on the central drainage core will reduce its thickness over time, and with it the capacity for water flow in the plane of the material. This is discussed separately in chapter 3.5.

To provide satisfactory design in these applications we need to establish:

- The time-dependent extension or creep under sustained load.
- The stress relaxation of a prestressed reinforcement.
- The time to failure under sustained load (creep-rupture).
- For seismic and similar applications, the change in strength and modulus following a period under sustained load.
- The dependence of the above on temperature.

The effects of influences other than load are covered in separate chapters.

## 3.4.2 Creep strain

### 3.4.2.1 Measurement of creep strain

In a tensile test a specimen of geosynthetic is gripped and then pulled apart by the crosshead of a tensile testing machine moving at a constant, standardised, rate. This is known as ramp loading. Load is measured by a load cell in series with the grips, while extension is measured over the central part of the specimen, known as the gauge length, to exclude any slippage in the grips. The load is generally expressed as a percentage of the reference tensile strength, in order to make it easier to transfer measurements from one product to another in the same range. The reference tensile strength should be measured using the same grips as will be used for the creep testing, and on the same batch of material.

Load is plotted against strain as the solid line in figure 3.29. Test methods are defined by ISO 10319 and ASTM D 4595 and D 6637 (note the differences in applied strain rate). The inflexion at low loads is peculiar to polyester fibres and is due to a change in the relative positions of the side groups that takes place when the polymer chain is strained. It is known as the gauche-trans transition and is reversible.

Suppose that the geosynthetic specimen is loaded only as far as a fixed load, less than the rupture strength. The stress-strain curve will be interrupted. Because of creep, however, the specimen will continue to extend under the fixed load, tracing a horizontal line on the graph. The rate of extension will decrease with time.

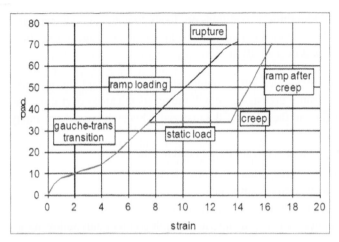

Fig. 3.29 The left hand line shows the stress-strain plot of a polyester geosynthetic under ramp loading. If loading is interrupted and the load held constant the geosynthetic will continue to extend as shown by the red line. After this period there may be a change in the modulus (see section 3.4.5).

Tensile creep for a geosynthetic is measured according to the standards ISO 13431 and ASTM D 5262, in which a specimen normally 200 mm wide is placed under a constant load for a set time, typically 1 000 hours (six weeks) or 10 000 hours (1.14 years), and the elongation monitored. Such tests can be performed over a range of loads and if required at various temperatures. Specimens should be tested in the direction in which the load will be applied in use, held by hydraulic or roller grips. The loads often amount to several tonnes and are therefore usually applied by weights operating through a lever. Where the loads are particularly high it may be possible to test narrower specimens, but edge effects, such as contraction of a nonwoven polypropylene or withdrawal of the weft fibres of a woven polyester, mean that these specimens may not behave in the same manner as wide ones. For this reason tests should be performed beforehand to demonstrate that the specimen width chosen is typical of the wide width geotextile. For geogrids this is not generally a problem. A small preload is applied to keep the specimen under tension at the start of loading and reduce the uncertainty in defining zero strain, particularly for nonwoven materials. Elongation is measured between points separate from the grips, as in tensile testing. The extensometers used to measure extension include linear variable differential transformers (LVDTs), potentiometer displacement devices, non-contact laser measurement, digital image analysis and, for slow creep, manually read dial gauges. A typical creep test is shown in progress in figure 3.30.

Creep is always tested in air. In real applications part of the load will be carried by the soil, reducing the load on the geosynthetic, but this cannot be simulated reproducibly in the laboratory. Tests in air are reproducible and, as the strain in service will be less, the measurements represent the worst-case situation (i.e. result in conservatively safe designs).

Practically all creep measurements are made in one direction only, since in most reinforcing applications the load is effectively uniaxial. Biaxial testing is particularly complex. In geogrids the creep under biaxial loading would not be expected to differ from that measured in each direct independently. In woven materials there would be a greater effect due to the effect of crimp in the weave (Greenwood 1990) while the creep strain of nonwoven materials under biaxial loading is likely to be considerably less that that measured uniaxially. ISO 13431 takes this at least partly into account by the use of suitably wide specimens to limit the effect of lateral contraction.

Typical loads to be used for creep strain measurement are 10%, 20%, 30%, 40% and possibly 50% and 60% of tensile strength, though the latter err into the region of creep-rupture. The minimum duration of index tests should be 1000 h (6 weeks), while tests intended for use in design should last up to one year (8760 h) or 10000 h (14 months). The standard test temperature is 20 °C and the relative humidity 65%.

Fig. 3.30 Creep testing of a geogrid, using roller (capstan) grips protected with white nonwoven material. Extension is measured between separate horizontal aluminium bars using LVDT extensometers. Load is applied by means of calibrated weights through a lever mechanism. The temperature and humidity control equipment is behind the creep rig. Photo: ERA Technology Ltd.

### 3.4.2.2 Presentation of creep strain

Creep diagrams show strain plotted against time. The strain is calculated as the elongation divided by initial gauge length, as in a tensile test. Time is generally quoted in hours and plotted either linearly or as a logarithm, which makes it easier – some would say too easy – to predict the behaviour after many years. Note that 1 000 000 h equals 114 years and is often used as a convenient figure for long-term lifetime of a geosynthetic. The data should also be tabulated.

Loads are generally expressed as a percentage of the reference tensile strength taken from rapid constant rate of strain tests described earlier, in order to make it easier to transfer measurements from one product to another in the same range. The reference tensile strength should be measured using the same grips as will be used for the creep testing, and on the same sample of material. To minimise further the effect of material variability the specimens used for measuring both strength and creep may be taken from the same ribs or fibres in the direction of reinforcement. It is normal to perform creep tests over a range of loads as described in section 3.4.2.1 and to plot these on a single diagram. Figure 3.31 shows typical creep curves for a polyester geosynthetic.

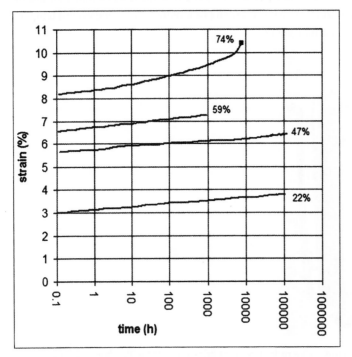

Fig. 3.31 Typical creep curves for a polyester geosynthetic plotted as strain against the logarithm of time (Greenwood et al 2000). At 74% of the reference tensile strength the specimen ruptured after 7785 h, or 10½ months.

Creep curves are sometimes divided into three regions: primary, secondary and tertiary. Primary creep strains are characteristically linear when plotted against a logarithmic time scale, meaning that the rate of creep decreases with time. Secondary creep strains are typically linear when plotted against an arithmetic time scale, meaning that the rate of creep is constant. Tertiary creep is the rupture phase of creep and is characterized by a rapidly increasing creep rate with time. In the primary creep of nonwovens and, to a lesser extent, woven materials, rearrangement of the structure of the geosynthetic structure contributes to primary creep. Elsewhere creep is governed by the creep of the polymer. Polyolefins (HDPE

206

and PP) tend to exhibit all three stages, depending on the load level, whereas PET tends to only exhibit primary and tertiary creep.

This nomenclature has been inherited from the testing of high temperature metals and refers to plots of strain time on a linear scale. On a logarithmic scale, as generally used for geosynthetics, the appearance of the three regions is completely different (figures 3.32 and 3.33).

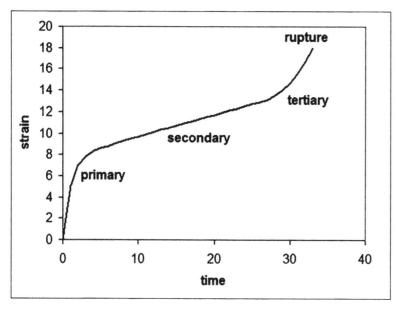

Fig. 3.32 Schematic diagram of strain plotted against time showing the appearance of primary, secondary and tertiary creep.

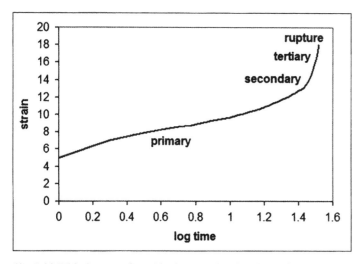

Fig. 3.33 With time on a logarithmic scale showing the difference in appearance of primary, secondary and tertiary creep.

In most applications of geosynthetics creep is in the primary stage. It is essential that it does not extent into the tertiary stage heralding the approach of rupture.

### 3.4.2.3   Typical creep curves

Figure 3.31 shows four typical creep curves for a polyester geosynthetic plotted as strain against time, with time on a logarithmic scale. The loads are indicated as percentages of the reference tensile strength. At 22 % of tensile strength the geosynthetic extends by 3 % on loading and then creeps at an ever decreasing rate. Plotted against the logarithm of time, however, this appears almost as a straight line. This is typical of polyester reinforced geosynthetics, where the extension on loading is proportionally high while the creep after loading is low. In the region of the gauche-trans transition described in section 3.4.2.1 the initial strain can be very sensitive to applied load and the creep rate can increase for a time until the transition is complete. Low creep rates are also observed in polyvinyl alcohol and stiff aramid fibres.

Figure 3.34 shows creep curves for a woven polypropylene, for which the initial strains are lower the rates of creep are higher, changing from primary towards secondary and tertiary. Figure 3.35 shows curves for an early polyethylene geogrid which show a similar effect.

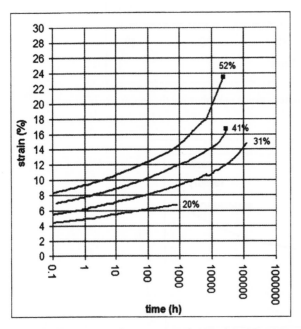

Fig. 3.34 Creep curves for a woven polypropylene (Greenwood et al. 2000).

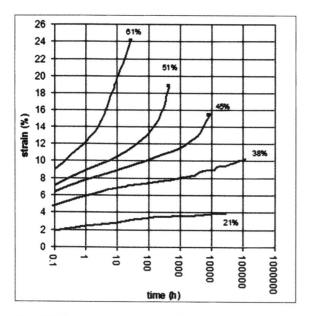

Fig. 3.35 Creep curves for a polyethylene geogrid (Greenwood et al. 2000, corrected).

Except where deliberately raised for the purpose of creep acceleration, the temperature of testing should be 20°C as stated in ISO 13431. All polymeric materials creep faster at higher temperatures, and particularly polyolefins (polypropylenes and polyethylenes) which are above their glass transition temperature, $\theta_g$ (see section 3.4.8). One result of this is that due to the nonlinear nature of the acceleration the additional elongation that takes place on a single hot day will far exceed the reduction in elongation on a correspondingly cold day. Another is that variations in temperature are the single greatest contributor to the uncertainty of geosynthetic creep strain measurement.

### 3.4.2.4  Ramp-and-hold tests

A frequent problem when presenting creep data from a range of loads is the variability in the strain on loading. This is particularly common in polyester reinforcements, where the initial strain appears to be highly sensitive to the method of loading and clamp arrangement. The creep strain after loading is comparatively reproducible. Isochronous curves generated from data with variable initial strains appear irregular and kinked. To reduce this effect for a long-term creep test replicate, short-term creep tests called 'ramp-and-hold' tests are performed at the same load, each lasting no more than an hour. The strains at the end of this period, together with the strain of the long-term creep test after the same period, are then averaged. The long-term creep data is then shifted up or down enabling the long-term creep test plot to pass through this average strain at the same set time. In figure 3.36 a long-term creep test has been performed together with two additional tests lasting one hour. The data from the long-term creep tests has then been shifted upwards to pass through the average of the three strains after one hour. Additional ramp-and hold tests will improve the accuracy of the average.

In using the data it must be remembered that strains measured from the start of installation may be subject to this variability in full, since the design engineer has no control over how the geosynthetic is to be loaded. Strains measured from the completion of construction are not subject to this variability.

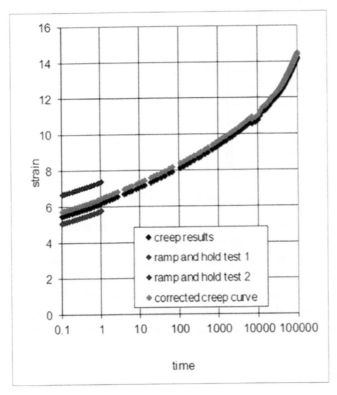

Fig. 3.36 Creep curve corrected for ramp-and-hold tests. The black line shows the original creep-curve. Two supplementary tests (blue, brown) were performed under identical conditions but lasting only one hour. The average of the three creep strains after 1 hour was calculated and the creep curve shifted vertically to pass through this average value (red).

### 3.4.2.5 Sherby-Dorn plots

A Sherby-Dorn plot is a form of presentation borrowed from the creep of high temperature metals. In it the rate of creep, on a logarithmic scale, is plotted against strain on a linear scale. An example is shown in figure 3.37. The rate of creep should be measured over an interval long enough to minimise scatter. Curves that are linear or concave downward indicate that only primary creep is occurring and that no rupture is likely. Curves that are parallel to the strain (x) axis indicate secondary creep: in figure 3.37 the creep rate appears to level out at a constant minimum of between $10^{-6}$ and $10^{-7}$ h$^{-1}$. Another example is shown in figure 3.38.

Although not shown here, a curve that is concave upwards would indicate that tertiary creep is occurring, and that rupture is likely. The plot can then be used to predict rupture when no actual rupture is observed.

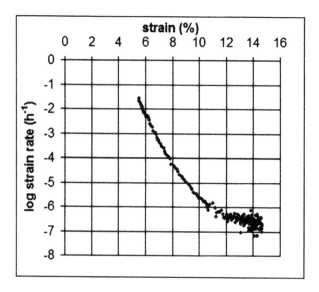

Fig. 3.37 Sherby-Dorn plot of strain rate plotted against strain for a polyethylene geogrid. Above 12% strain the rate of strain is approximately 10-6.5 h-1 (3 × 10-7 or 3 × 10-5 % per hour) and appears to be approaching a lower limit (Greenwood et al. 2000).

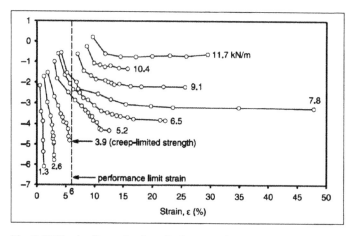

Fig. 3.38 Sherby-Dorn plot from Bathurst et al. (2006). The horizontal sections of the curves at higher loads and strains indicate secondary creep. In this case a strain limit of 6% is set, corresponding to the limit of primary creep.

### 3.4.2.6   Predicting creep strain by extrapolation

The design of a reinforced soil structure may include a limit on the acceptable total strain during the entire design life, or else on the amount of additional strain following construction. In the design this is called the serviceability limit state. This strain will depend on the load applied to the reinforcement by the weight of the soil plus the weight of any other surface load (e.g. traffic).

Creep strains are measured over much shorter durations than the expected design life. To make predictions for these lifetimes, creep curves can be extended manually, numerically or by time-temperature shifting. There is no fundamental law relating creep strain with time; most theoretical work on the subject has described the dependence of creep on temperature or physical ageing, or secondary creep, while leaving the shape of the primary creep curve itself to be fitted empirically. What appears right is still a matter for subjective judgment.

The equation most commonly used to extrapolate creep strain is the power law, which for creep appears as:

$$\varepsilon = A + B\, t^n$$

where $\varepsilon$ is strain, t is time and A, B and n are constants to be optimised by use of a routine such as Excel Solver. The power law can also be fitted to strain-time data using the Excel TREND function. Alternatively, if a plot of creep against the logarithm of time is clearly linear, then one can write

$$\varepsilon = C + D \log t$$

where C and D are constants.

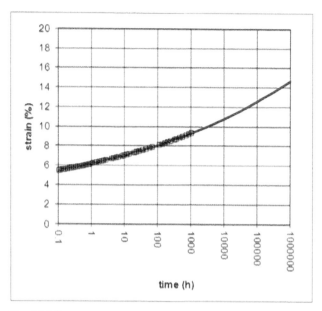

Fig. 3.39 Creep strain measured over 1 000 h for a woven polypropylene. A fitted curve has been calculated on the basis of a power law and extrapolated to $10^6$ h.

Figure 3.39 shows a power law fitted to creep measurements lasting 1 000 h on a woven polypropylene. A power law has been fitted:

$$\varepsilon = 1.78 + 4.44\ t^{0.0774}$$

The measurements cover a 1 000 h (six week) period. The fit is visibly good. In figure 3.39 the power law has been used to extrapolate the curve to 1 000 000 h (114 years); after 100 000 h predicted strain is 12.6% and after 1 000 000 h, 14.7%. To the eye, the prediction appears credible.

In reality, this test was continued for 100 000 h (11 years). In figure 3.40 the power law prediction from the first 1000 h is compared with the measurements between 1 000 h and 10 000 h. The agreement is good, validating the extrapolation by a factor of ten.

In figure 3.41 the power law prediction from the first 1 000 h is compared with the measurements between 10 000 h and 100 000 h. The measured creep strain of 14.3% exceeds the predicted value of 12.6%. This is an example (but not a proof) of the general rule of thumb that creep tests on polymers may be extrapolated by up to a factor of 10, for example from 1 000 to 10 000 h, but beyond that point a precautionary factor should be included. More correctly, any prediction should consider the transition from primary to secondary and tertiary creep as shown in figure 3.38. In ISO TR 20432 this is taken into account by a safety factor (see section 3.7.2.2). No safety factor is specified for creep strain, but the level of uncertainty should be regarded as similar.

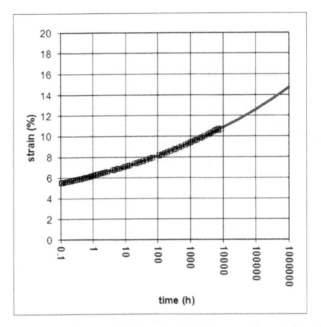

Fig. 3.40 As figure 3.38 but with the measurements extended to 10 000 h. The fitted curve is based on measurements taken up to 1 000 h.

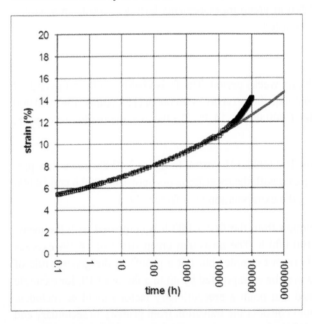

Fig. 3.41 As figure 3.38, but with the measurements extended to 100 000 h. The fitted curve is that based on the measurements to 1 000 h.

The following advice applies specifically to the extrapolation of creep strain:
- Keep theories simple.
- Do not extrapolate by more than a factor of 10, or if this is unavoidable consider applying a safety factor.
- If possible predict the transition from primary to secondary creep.
- Perform ramp-and-hold tests to reduce the variability in initial strain.
- Do not be complacent about logarithmic scales: they can make extrapolation look easier than it is.
- Do not trust computer-generated functions without knowing the basis for them. There is no "right" function to describe creep strain. You are the final judge of what is acceptable.
- Use the thermal shift factors to predict strains at different reference temperatures (3.4.6).
- Plot the data as isochronous curves (section 3.4.2.10).

### 3.4.2.7    Prediction by time-temperature shifting

Extrapolation by no more than a factor of 10 means that a design life of 100 years will require creep measurement lasting 10 years. Since few people have the patience to wait this long, such data are rare. In this case accelerated tests will have to be considered, particularly those in which creep is accelerated by raising the temperature.

Temperature acceleration of creep curves uses the established method of time-temperature superposition of the creep of polymers without direct reference to Arrhenius' formula. The procedure is as follows:

- Perform a series of separate creep tests under the same load but at different temperatures, with intervals generally not exceeding 10 °C.
- Plot the resulting creep curves on the same diagram as strain against log $t$.
- Taking the creep curve at the lowest temperature as the reference temperature, shift the creep curves at the higher temperatures along the time axis until they coincide. The curves should overlap.
- The resulting single "master" curve is the predicted long-term creep curve for the reference temperature.

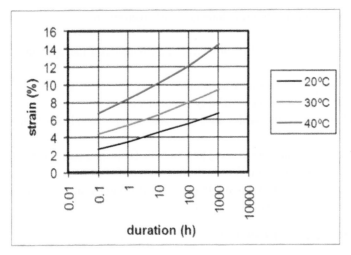

Fig. 3.42 Creep strain measurements performed on similar specimens under the same load at 20 °C, 30 °C and 40 °C.

Figure 3.42 shows typical results of creep strain tests on specimens of the same polyethylene geogrid at the same load and temperatures of 20 °C, 30 °C and 40 °C. In figure 3.43 the results at 30 °C and 40 °C tests have been shifted along the time axis until they form an extension to the results at 20 °C. The time has been extended from 1000 h to over 1000000 h at which point the predicted strain is 12%. The shift factors are expressed in two ways. At is the amount by which the times are multiplied. at, which is more commonly used, is equal to log At and is the amount by which each curve is shifted along the logarithmic axis. at should be plotted against temperature where it should form a straight line or smooth curve passing through zero at the reference temperature (figure 3.43).

Temperature steps of 40 °C and 60 °C are more appropriate for polyesters.

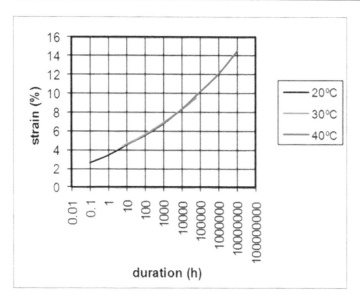

Fig. 3.43 As figure 3.42, but with the creep curves for 20 °C, 30 °C and 40 °C shifted along the time axis to form a continuous master curve, which provides an extension to the creep curve at 20 °C.

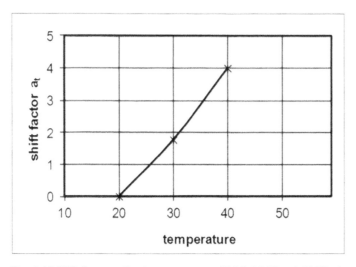

Fig. 3.44 Shift factor at for the temperatures 20 °C, 30 °C and 40 °C taken from figure 3.43 and plotted against temperature to ensure that they lie on a straight line or smooth curve.

If this is done by computer using an optimization utility such as SOLVER in Excel, proceed as follow:

- Assume a power law equation ($\varepsilon = A + B\,t^n$) for the creep curve at 20 °C
- Estimate two shift factors $a_{30}$ and $a_{40}$ for the horizontal displacement of the 30 °C and 40 °C curves, respectively.
- Select trial values for $a_{30}$ and $a_{40}$.

217

- Increase the values of log $t$ for the results at 30 °C and 40 °C by adding the appropriate shift factor.
- Select trial values for $A$, $B$ and $n$ in the power law equation and calculate the predicted strains.
- Tabulate the predicted strains and the actual strain measurements at each point.
- Calculate the square of each difference and add them to give the sum of squares of the errors.
- Minimize the sum of squares by varying the five constants $A$, $B$, $n$, $a_{30}$ and $a_{40}$.
- Check that the predicted curve is reasonable: you are the ultimate judge of what is acceptable, not the computer.

When this method is applied to moulded thermoplastics, a correction may have to be made for physical ageing, but this can be ignored for oriented polymers as used in reinforcing geosynthetics. A more serious problem with geosynthetics, particularly polyesters, is the variation in initial strain mentioned earlier. When performing time-temperature shifting it is strongly advisable to perform ramp-and-hold tests to reduce the errors in initial strain. Alternatively the stepped isothermal method can be applied.

### 3.4.2.8   The stepped isothermal method (SIM)

The stepped isothermal method, or SIM, ingeniously avoids the problem of variability in initial strain by using a single specimen. The method has now been standardised as ASTM D6992:2003 to which reference should be made for a full description. The general procedure is as follows:

- Set up a creep test in the normal way, but within an oven capable of raising the temperature within one minute and then maintaining the new temperature to within ±1 °C. This will require a good quality thermal controller and a data logger capable of monitoring strain, load and temperature for a period of typically 24 h. Strain has to be recorded twice a second during the initial loading ramp periods and twice a minute during the constant load periods.
- Apply load and allow the test to continue for approximately 3 h.
- Increase the temperature and then maintain it to within the required tolerance: a step of 7-10 °C has proved satisfactory for polyethylene and polypropylene while 14 °C is appropriate for polyester.
- Increase the temperature further in equal steps. The duration at each step should ideally be the same, but longer steps are acceptable, for example overnight.
- Stop at the maximum acceptable temperature: approximately 60 °C for polyethylene, 70 °C for polypropylene and 90 °C for polyester and similar fibres (i.e. below the glass transition temperature unless otherwise validated).
- Tabulate the results on a spreadsheet and plot as a creep diagram (figure 3.45).

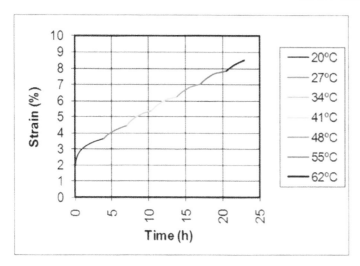

Fig. 3.45 Creep data from a single constant load SIM test, plotted as strain against time.

- Calculate creep modulus, which is load divided by strain.
- Calculate the time from the beginning of each step. This point can be taken as the time at which the new temperature is within 1 °C of the target.
- Plot creep modulus against the logarithm of the time from the beginning of each step (log $t$ ) (figure 3.46).

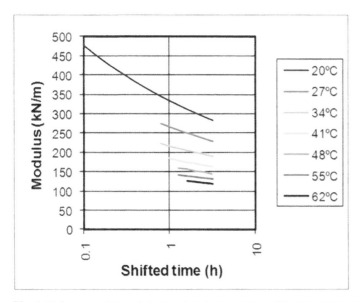

Fig. 3.46 Creep modulus plotted against the logarithm of the time following each temperature change individually.

- Tabulate the following shift factors for each temperature:
    - Thermal history factor $t'$, to be added to all values of $t$ for that temperature.
    - Thermal shift factor, to be added to log $(t + t')$ for all times at that temperature (equivalent to at)
    - Vertical shift factor, to be added to all strains for that temperature.
- Having included these corrections, plot the adjusted values of creep modulus against the shifted values of log $t$ (figure 3.47).

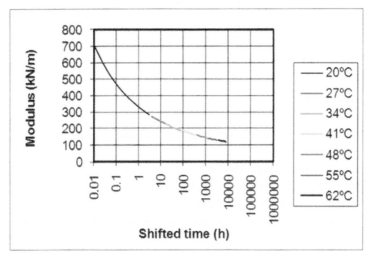

Fig. 3.47 The sections of creep modulus curve in figure 3.46 have been corrected using the vertical and thermal history factors and then shifted along the time axis (by the thermal shift factors) to yield one continuous curve.

- Optimise the correction factors to yield a smooth master curve with the sections measured at the individual temperatures lined up nose-to-tail with minimal discontinuities in the strain and gradient. In SIM the sections of creep curve at each temperature do not overlap.
- Calculate the strain from the shifted creep modulus and plot it against log $t$ (figure 3.48). This is the predicted master curve for a load applied at the reference temperature.
- To confirm the validity of the process, plot the shift factor against temperature and ensure that it follows a straight line or smooth curve passing through zero at the reference temperature (figure 3.49). If there is a discontinuity, the shifting procedure is invalid. Thermal shift factors for oriented fibres tend to lie between 0.08 and 0.14 per °C (Lothspeich and Thornton 2000). This curve is also used if the design temperature differs from the reference temperature (see section 3.4.6).
- Read off the predicted strain at the design life.

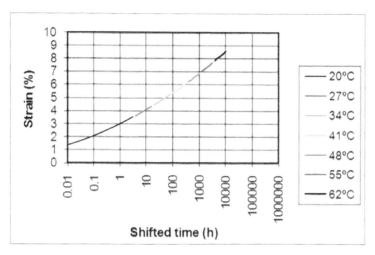

Fig. 3.48 Predicted creep curve derived from a single constant load SIM test.

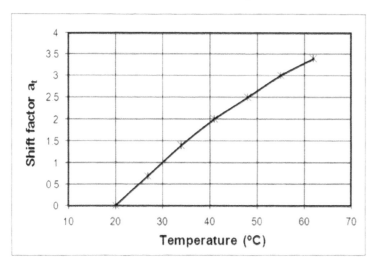

Fig. 3.49 Horizontal shift factor at plotted against temperature. The absence of any discontinuity demonstrates that the creep mechanism is the same at all test temperatures.

If the activation energy is required, for example to understand the physical mechanism governing the creep, then it may be calculated from the formula

$$a_T = E_a / 2.3036R \, (1/\Theta_1 - 1/\Theta_2)$$

where $a_T$ is the shift of the creep curve (plotted against the logarithm of time log t) between the absolute temperatures $\Theta_1$ and $\Theta_2$. $E_a$ is the activation energy in J/mol and R is the universal constant 8.314 J/mol.K. The numerical factor 2.3036 is necessary to change to natural logarithms.

221

The high values of shift factor lead to a remarkably high degree of acceleration that assists with long-term prediction: a temperature rise of 40 °C results in a shift factor of about 3.3. Time is then multiplied by $10^{3.3}$ or approximately 2000. A design life of 100 years will require a test duration of two and a half weeks.

### 3.4.2.9    Validity of SIM

SIM has proved a powerful method for providing accelerated creep data. The shift factors for oriented fibres are particularly high, typically between 0.08 and 0.12 decades per °C. For polyester fibres the temperature can be raised as high as 90 °C, since it has been shown that the crystalline regions and tie molecules which govern the mechanical behaviour are unaffected by the transition in the amorphous material at 70 °C or below. It thus becomes possible to simulate the entire lifetime of a reinforcing geosynthetic in an overnight test.

The validity of SIM to simulate real creep behaviour has been demonstrated for three out of four materials by comparison with 12 year tests (Greenwood et al. 2004). It should never be forgotten, however, that this is a highly accelerated test and that the critical conditions on the validity of accelerated testing apply.

SIM is valid for extruded polyethylene grids but of less interest, since time-temperature shifting can be applied equally to these materials where there is less variability in initial strain, and the temperature cannot be raised above 60 °C.

Yeo and Hsuan (2008) introduced a modification to the SIM procedure. Instead of introducing correction factors t' they plot strain from the time at each temperature step took place and then eliminate the section of the creep curve for that temperature that is non-equilibrium, i.e. has a different slope. The master curve then has short gaps at each temperature step, and the vertical shifts are used to bring the different sections into line. This procedure was shown to give improved results for HDPE geogrids at higher strains, as judged by agreement with conventional time-temperature shifting and by the activation energies, and is recommended for use with polymers whose creep behaviour is unknown. For PET geogrids and for HDPE at lower strains the procedure gave results that were identical to those derived using the procedure of ASTM D6992.

Kongkitkul et al (2007a) have modified their three component model to allow for the effects of temperature. They predict that creep-rupture curves generated using SIM are similar at shorter lifetimes but give longer results from that using the conventional approach at very long times.

### 3.4.2.10  Presentation of creep data: isochronous curves

Design codes for reinforced soil may specify a strain limit, such as a maximum total strain of 5% or a maximum strain after construction of 2%. These criteria have then to be converted to a maximum acceptable load on the geosynthetic and the quantity or grade of geosynthetic required. For this purpose the most convenient means of depicting the creep data is an isochronous plot which comprises an array of load-strain curves, similar to the one obtained from a tensile test, but with each curve representing a different duration. Given a set of predicted creep curves for the same material at different loads, for each load the strains are read off at chosen durations such as 1 hour, 10 hours, 100 hours, up to the design life. These points are then plotted as load against strain (figure 3.50). The points corresponding to each elapsed time are then joined. These isochronous curves or isochrones pass through the origin and should be similar in shape to the original reference load-strain curve but lie to the right of it.

For a maximum acceptable strain of 5% and a lifetime of 100 000 h (black curve) figure 3.50 shows the maximum acceptable load for this geosynthetic to be 6 kN/m. Note that when testing geosynthetics strains are measured from a set preload (defined in ISO 10319 and ISO 13431 as 1 % of the reference tensile strength) and that some woven and particularly nonwoven materials may exhibit considerable irreversible strains below this initial loading.

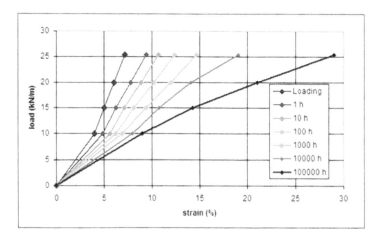

Fig. 3.50 Isochronous diagram for a geosynthetic. The curve marked 'Loading' refers to the reference load-strain curve.

If the criterion is for a maximum acceptable strain following construction ('time-dependent' or 'in-service' strain) then the duration of construction must be defined. A creep test is loaded within less than a minute while on site the maximum load of a geosynthetic may not be realised for a month after backfilling commences. As an example, if the maximum acceptable 'in-service' strain after 100 000 h is 2%, and 'service' is taken as commencing after just 100 hours (4 days), the maximum acceptable load from figure 3.50 would be that at which the isochronous lines for these two durations (light green and black curves) are separated by no more than 2% strain. This is at a load of approximately 5 kN/m.

It is clear that during construction the rate of loading is much slower than in a tensile test. This method assumes that the strain achieved by rapid loading followed by a period at constant load is the same as if that same load had been applied slowly over the same period. To define this more precisely, tests would need to be performed at very slow rates of strain followed by periods a constant load.

### 3.4.3   Creep-rupture

#### 3.4.3.1   Residual strength

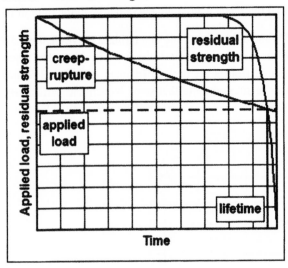

Fig. 3.51 Schematic diagram showing the fall-off in residual strength with time when a sustained load is applied to a geosynthetic. Rupture occurs when the residual strength equals the applied load. The creep-rupture curve is the locus joining creep-rupture points corresponding to different applied loads.

Under sustained load the strength of a geosynthetic gradually decreases. At the moment when the strength of the specimen has fallen to the point where it equals the applied load, the specimen breaks. Up to that moment its strength will have been greater. Measurements of this 'residual' strength of samples under sustained load show that they retain their original 100% strength over the majority of their lifetime (Voskamp et al 2001b). Under sustained load, apart from any environmental or installation damage effects, strength falls off during a relatively short period before final failure. Sustained load is thus a Mode 3 form of degradation (figure 3.5.1). The material effects that lead to this relatively sudden fall-off in strength are not understood for all polymers. A model to describe the creep behaviour of PET and the sudden fall-off in strength has been developed by Voskamp et al (2001a, 2001b, 2006). It describes the changes in the PET fibre which take place at molecular chain level during continuous, long term loading.

Residual strength is important in applications where the function of the geosynthetic is to withstand a sudden excess load, such as in seismic design. If the permanent or sustained load in the geosynthetic is low compared with the available strength, it may be possible to assume that the strength of the geosynthetic is unaffected by the applied load throughout the active life of the reinforcement. It may, however, decrease due to environmental or other effects.

224

### 3.4.3.2 Measurement of creep-rupture

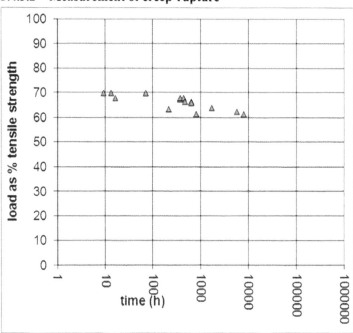

Fig. 3.52 Creep-rupture data with applied load plotted against the logarithm of time to rupture. Each point represents one test taken to rupture.

Under sustained load there will be a lifetime at which the material ruptures under the effect of load alone and for a prescribed design life it is essential that the load is low enough to exclude any possibility of rupture. To predict this lifetime requires a creep-rupture curve, in which the applied load, generally expressed as a percentage of reference tensile strength, is plotted against the logarithm of time to failure for that load (figure 3.52).

A creep-rupture diagram is the result of a number of separate tests, each of which yields one point on the diagram. Each point represents the moment at which, in an individual test, a specimen has broken. It is important to recognize that a creep-rupture diagram does not depict the loss of strength of a geosynthetic with time under load, though it may appear to do so. It is simply the locus of points each representing the lifetime of a specimen under a particular load.

In view of the scatter of the data – at a single load the time to rupture may vary by as much as a factor of ten – ISO/TR 20432 recommends that at least twelve tests are performed. There should be at least four different loads. For conventional testing four of the test results should have rupture times between 100 hours and 1 000 hours, and four of the test results should have rupture times of 1 000 hours to 10 000 hours, with at least one additional test result having a rupture time of approximately 10 000 hours (1.14 years) or more. Since the times to failure can be extremely unpredictable, only nine of the twelve rupture times are specified as having to lie within particular limits. Start testing at the highest loads in order to estimate the approximate form of the creep-rupture diagram, and only then set up the tests intended to last for 10 000 h or more. Record the time to rupture electronically, although daily visual observation may suffice for longer tests. Observe and record the nature of the failures.

Although it is customary to measure creep strain in the course of the test, this information does not generally help in predicting the time to rupture.

Tabulate the creep rupture data for the product as load, expressed as a percentage of the batch reference tensile strength, and time to rupture in hours. Include incomplete tests and tests terminated before rupture ('run-outs'), with the test duration replacing the time to rupture, with a note that they were not completed.

If products from a range are regarded as sufficiently similar then it may be appropriate to plot the creep-rupture points from all products on one diagram, each one expressed as a percentage of its reference tensile strength. Experience in North America is that that these creep rupture curves show no more error than the inherent scatter in the virgin strength of the materials (Bathurst et al. 2012). In the United Kingdom one product of a range has been taken as the reference material with up to 12 creep-rupture points, to which data from the products at the extreme ends of the range are added for comparison. However, in some cases the differences have been significant.

### 3.4.3.3   Presentation of creep-rupture data

Creep rupture data should be plotted as load ( $T$ ), expressed as a percentage of the batch reference tensile strength ( $T_B$ ), against the logarithm of the time to rupture in hours (log $t_R$ ) (figure 3.53). This is referred to as a semi-logarithmic plot. Assuming that the points lie visibly on a straight line, the best fit straight line, known as the regression line, is fitted as follows. $y = T / T_B$ or a function of $T / T_B$ and $x = \log t_R$.
- Calculate $x_{mean}$, the mean of all values of $x$, and $y_{mean}$, the mean of all values of $y$, excluding any incomplete tests ('run-outs')
- Calculate $T_{xx} = \Sigma( x - x_{mean} )^2$, $T_{yy} = \Sigma( y - y_{mean} )^2$ and $T_{xy} = \Sigma( x - x_{mean} ) ( y - y_{mean} )$, where the summation is over all points
- Equation of regression line: $y = y_{mean} + ( T_{yy} / T_{xy} )( x - x_{mean} )$
- In terms of the original data, $y = a - b \log t_R$, , $a = y_{mean} - x_{mean} T_{yy} / T_{xy}$ and $b = T_{yy} / T_{xy}$.

Figure 3.53 shows the creep-rupture data with the regression line added.

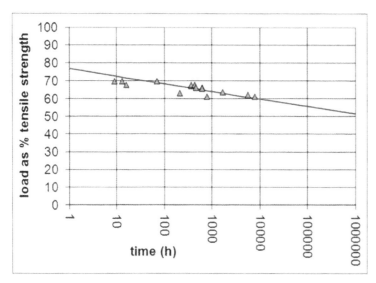

Fig. 3.53 Creep-rupture data with regression line added.

The gradient of the graph equals ( $T_{yy} / T_{xy}$ ), should be negative, and is expressed in terms of percentage of tensile strength per decade of time (one decade equals the interval between log 10 and log 100, or log 100 and log 1 000). Typical values for polyester are about -3%/ decade, indicating that if the load is reduced by just 3% of the tensile strength the lifetime will increase tenfold. The intercept $y_0$ on the line $x = 0$ (i.e. at log $t = 0$; $t = 1$ h) is given by $y_0 = y_{mean} - x_{mean} T_{yy} / T_{xy}$.

Note that in contrast to most scientific plots, the independent variable (here the applied load) is plotted on the $y$ axis while the dependent variable (here the measured time) is plotted on the $x$ axis. The formulae that follow therefore differ from those conventionally found by having x and y interchanged. If using Excel Slope and Intercept functions, enter $x = T / T_B$ (or function of $T / T_B$ ) and $y = $ log $t_R$ and plot the graph as log $t_R$ against load.

If there are incomplete tests, repeat the calculation adding the incomplete tests one by one. If their inclusion causes σ to increase, retain the test in the calculation. In general such points lie to the right of or above the regression line.

Record the duration of the longest test that has ended in rupture, or the duration of the longest incomplete test whose duration has been included in the regression calculation: this duration is denoted as $t_{max}$.

As an exercise and a check, the data used for the diagrams above is:

| Load as percentage of tensile strength (%) | Time to rupture $t_r$ (h) | log $t_r$ |
|---|---|---|
| 70.0 | 9 | 0.9542 |
| 70.0 | 13 | 1.1139 |
| 68.0 | 16 | 1.2041 |
| 70.0 | 70 | 1.8451 |
| 63.5 | 210 | 2.3222 |
| 67.5 | 363 | 2.5599 |
| 68.0 | 365 | 2.5623 |
| 68.0 | 435 | 2.6385 |
| 66.5 | 463 | 2.6656 |
| 66.5 | 627 | 2.7973 |
| 66.0 | 625 | 2.7959 |
| 61.5 | 790 | 2.8976 |
| 64.0 | 1 670 | 3.2227 |
| 62.5 | 5 605 | 3.7486 |
| 61.5 | 7 791 | 3.8916 |
| $y_{mean}$ = | | $x_{mean}$ = |

$T_{xx} = 10.85$

$T_{yy} = 124.9$

$T_{xy} = -29.57$

Gradient $T_{yy} / T_{xy} = -4.225$

Intercept at $t_r = 1$ [log $t_r = 0$] = 76.7.

The straight line is then extrapolated to predict the maximum sustained load which can be applied to the geosynthetic without it breaking before the end of its design life. In figure 3.53 the load corresponding to a design life of 1000000 h (114 years) is 51.4% of the tensile strength. The reduction factor RFCR is the inverse (1/51.4% = 1.95).

If the plot is visibly nonlinear, a curve can be fitted instead of a straight line. Ideally the curve should be based on a physical model of the polymer, but there is no general model available. Conversely, load can be replaced by a mathematical function of applied load to achieve a linear plot: for example the function $y = \log ( T / T_B )$, resulting in a double logarithmic plot, has been shown to yield a straight line for polyethylene and polypropylene reinforcements. Use of a straight line plot assists with subsequent statistical analysis (Bathurst et al 2012).

A condition on the extrapolation is that there is no evidence or reason to believe that the rupture behaviour will change over this duration. It should be checked that at long durations, and at elevated temperatures, if used (see below):

- there is no abrupt change in the gradient of the creep rupture curve;
- there is no abrupt change in the strain to failure;
- there is no significant change in the appearance of the fracture surface of ruptured specimens.

Any evidence of an abrupt change, particularly in accelerated tests, invalidates the extrapolation unless it is justified separately. Particular attention is drawn to the ductile-brittle transition in un-oriented thermoplastics under sustained load such as is observed in polyethylene pipes under pressure; if this is likely to occur, reference should be the extrapolation methods of ISO 9080:2003.

#### 3.4.3.4   Lower confidence limit

BS 8006: 1996 (superseded) required the calculation of the lower confidence limit. This is not regarded as representative of the scatter, since the calculation reflects only the variability of the batch tested and not that of the entire production. The lower confidence limit is the lower branch of a hyperbola (the upper branch is the upper confidence limit) which is narrowest at the mean and fans out at longer times, representative of the greater degree of uncertainty (figure 3.54). The formula for this, where required, is

$$y = y_{mean} + (b/K)(x - x_{mean}) - (t\sigma_0/K)\sqrt{\{1 + 1/n + (x - x_{mean})^2/T_{xx}\}}$$

where

$b = T_{xy}/T_{xx}$

$K = b^2 - t^2\sigma_0{}^2/T_{yy}$

$n$ = number of readings

$T_{xx} = \Sigma(x - x_{mean})$

$T_{yy} = \Sigma(y - y_{mean})$

$T_{xy} = \Sigma(x - x_{mean})(y - y_{mean})$

$\sigma_0 = \sqrt{\{(Tyy - T_{xy}{}^2/T_{xx})/(n-2)\}}$

$t$ = Student's $t$ for $n - 2$ (one-sided, stated percentage e.g. 95%: this is equivalent to a two-sided probability of 90%. The percentages refer to the percentage of measurements above the lower limit and the percentage of points between the lower and upper limits respectively. The upper limit is calculated using the same formula for y with a positive sign before $t.\sigma_0$. If using Excel, Student's t is represented by the function TINV, followed by the two-sided probability level (expressed as the proportion of points outside the limit, thus 90% above is represented as 10% or 0.1) and the number of degrees of freedom. For the above example one would enter TINV(0.1,n − 2).

Note that this formula takes into account the reversal of *x* and *y* as explained in section 3.4.3.3.

The scatter of the entire production is taken into account by using a characteristic strength to calculate design loads in kN/m. The may be based on factory production statistics, or the nominal value and tolerance quoted for CE marking. Bathurst et al (2012) confirmed this by showing that the coefficient of variation of the ratio of measured to predicted geogrid creep-rupture strengths at a design life of 100 000 h was largely captured by (i.e. no greater than) the spread in reference tensile strength of these geosynthetic materials. This gives confidence that extrapolation of block temperature accelerated tests and SIM as described here is a reliable technique.

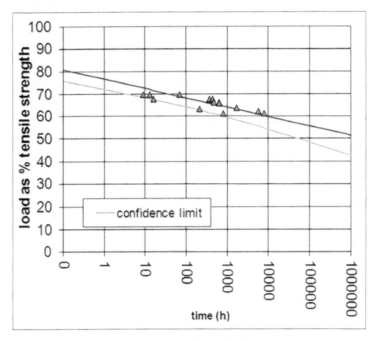

Fig. 3.54 Example of a 95% lower confidence limit.

### 3.4.3.5   Time-temperature shifting of creep-rupture data

Creep-rupture regression lines are extended to longer times and lower loads in order to yield a design load for the design life. If extrapolation is by more than a factor of ten, which it often is, an additional safety factor will be required. In order to improve the prediction, time-temperature shifting is used as for creep strain.

In conventional time-temperature shifting a number of creep-rupture tests are performed at each of a set of temperatures. The points are then shifted along the time axis to generate an extended master creep curve for the lowest, reference, temperature. This can be done in one of two ways.

In simple "block" shifting a single shift factor is assigned to each temperature. This factor is then added to the logarithm of each rupture time at that temperature (or, if shifting to a higher temperature, subtracted). The effect is to shift all points by the same amount, as a "block". The equation of the creep-rupture line becomes:

$$y = a - b\,(\log t_R + a_\theta)$$

where $a_\theta$ is the shift factor of the temperature $\theta$. $a$, $b$ and $a_\theta$ can then be calculated by optimisation methods (minimising the sum of the squares of differences between the calculated and measured values of $\log t_R$).

Figure 3.55 shows rupture points for three sets of tests performed at 20 °C, 40 °C and 60 °C respectively. These can be shifted along the time axis to give a single master curve for the reference temperature of 20°C, as shown in figure 3.56.

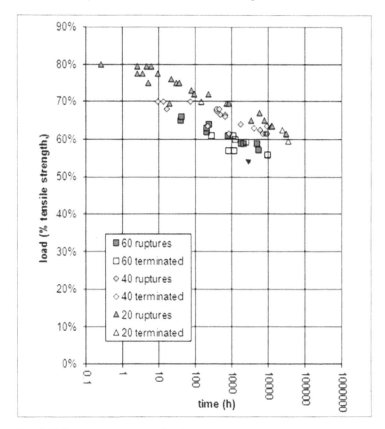

Fig. 3.55 Creep-rupture points for three temperatures.

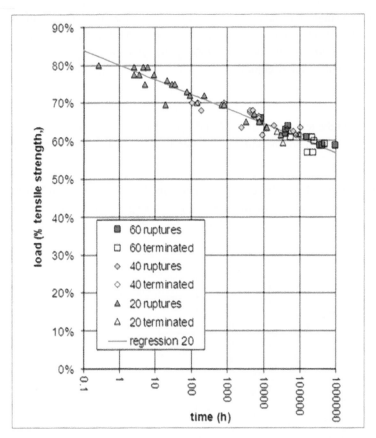

Fig. 3.56 Master diagram for 20 °C generated by shifting the points in figure 3.55 along the time axis. The design load for 1 000 000 h at 20 °C is 56.9% of reference tensile strength.

Temperature shift factors may also be derived by performing pairs of tests at identical loads (each expressed as a percentage of the local strength) but at different temperatures and comparing the times required to achieve either rupture or a set strain, the latter being usually the more precise (M. Dobie, personal communication).

Alternatively Zhurkov's equation can be used, also known as the Bueche-Zhurkov equation (Zhurkov 1965). This has the advantage of a scientific basis since it treats creep-rupture as a thermally activated mechanism modified by the presence of the applied load:

$$t_R = A_0 . \exp[(E-TV)/R\theta]$$

where $t_R$ is the time to rupture, T is the applied load, V is the coefficient of load with the units of volume, E is the activation energy, $\theta$ is temperature, R the universal constant 8.314 J/mol.K and $A_0$ is a constant as in Arrhenius' equation (see chapter 2.3). Taking logarithms of both sides,

232

$\ln t_R = \ln A_0 + E/R\theta - TV/R\theta$

Generate an optimal creep-rupture curve by inserting nominal values for $\ln A_0$, E and V, list the calculated and measured values of $\ln t_R$, calculate the squares of the differences between these values, sum the squares, and use a procedure such as Excel Solver to minimize this sum by finding appropriate values of $\ln A_0$, E and V. Figure 3.57 shows the optimized creep-rupture curves for the three temperatures, which converge lightly to the left. The design load for 20 °C is 54.7% of the reference tensile strength and $RF_{CR} = 1/54.7\% = 1.83$.

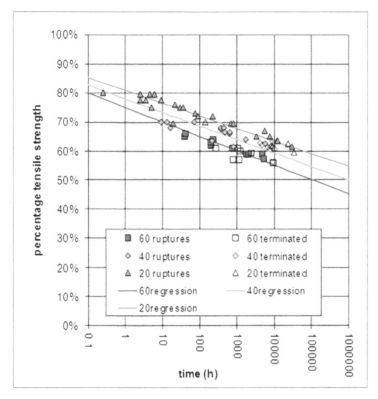

Fig. 3.57 Creep-rupture curves generated by fitting the points in figure 3.54 to Zhurkov's equation. The design load for 1 000 000 h at 20 °C is 54.9% of tensile strength.

Either method of time-temperature shifting can be used to derive the creep-rupture curve for a temperature other than the minimum test temperature, although Zhurkov's formula is the most convenient, requires no subjective shifting, and has the justification of being based on a simple physical model. The data in figure 3.57 give design loads for 1 000 000 h of 54.9 % of tensile strength for 20 °C, 52.3% for 30 °C, 49.9% for 40 °C and 45.2% for 60 °C.

SIM is also frequently used to generate creep-rupture points on an accelerated timescale. A test is performed exactly as described in section 3.4.2.7 but extended until rupture occurs. The shifted time is then calculated using the same procedure as for creep strain, in fact the measured rupture time depends entirely on correct shifting of the strain data. As in conventional testing, each test generates just one creep-rupture point.

ISO TR 20432 recommends the following rules for combining SIM and conventional test data that for the calculation of $RF_{CR}$. There should be at least six conventional tests, of which four should rupture between 100 and 10 000 h and one after more than 10 000 h, and six SIM tests, of which three should have rupture times between 1 000 and 100 000 h and three between 100 000 and 10 000 000 h. Regression lines should be derived for each method and $RF_{CR}$ calculated for times of 2 000 and 10 000 h. The two values of $RF_{CR}$ at 2 000 h may differ by no more than 0.15 and the same applies to two values of $RF_{CR}$ at 10 000 h. If these conditions are satisfied then all the results may be combined to give a regression line. If not, then SIM cannot be regarded as providing a valid acceleration for creep. An example database of $RF_{CR}$ values deduced from SIM testing reported in the open literature can be found in the paper by Bathurst et al (2012).

### 3.4.3.6   Alternative methods for measuring creep-rupture

Some US methods allow the design load to be derived by extrapolating the time to 10% strain instead of the time to rupture, thereby eliminating the need to wait until rupture occurs. Since the strains at rupture generally exceed 10% this will yield a conservative result, but makes the assumption that the strain to failure will not decrease with time to the point at which it is less than 10% for the required design life. While such a method would be unacceptable for isotropic polymers with a ductile-brittle transition, it is more than likely to be conservative for reinforcing geosynthetics.

Koo et al. (2008), in work based on a range of SIM tests, have used a 1% time-dependent strain instead of the 10% total strain to derive a long-term design strength. Similarly, for an extruded polyethylene geogrid Wrigley and Zheng (2008) have measured the duration after which there is a change in gradient of the strain/log time graph [$d\varepsilon/d(\log t)$] for each load. They have extrapolated the resulting load/log time characteristic in the same manner as for conventional creep-rupture.

McGown Khan and Kupec (2004a, 2004b, 2004c) address the problem of predicting the strain response to a history of variable loads and temperatures. They define an isochronous strain energy, being the area under the isochronous curve for a defined load, temperature and time, a measurement that is taken from relatively short term tests. This energy is then plotted against time and extrapolated to predict lifetimes under various end-of life criteria based on a limiting strain.

### 3.4.4 Stress-relaxation

Stress relaxation is the counterpart to creep. A geosynthetic will continue to creep only if the load is maintained and if the geosynthetic is free to extend. Thus the soil structure must extend with it. If the structure cannot move, for example if the geosynthetic is prestressed between two anchors, the geosynthetic keeps its original length and shape but the prestress relaxes. If the soil surrounding a reinforcement stops moving the prestress in the geosynthetic will relax.

If a sudden additional load occurs, for example an earthquake or an exceptional load, then the remaining prestress in the geosynthetic will prevent movement of the soil or anchor provided that the additional load does not exceed the prestress. As the prestress relaxes, so does the capacity of the geosynthetic to resist sudden loads diminishes.

Figure 3.58 shows the load-strain curve of a geosynthetic and, superimposed, the load-strain curve of a test which has been interrupted and allowed to relax with time.

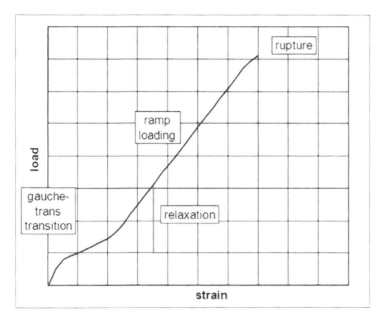

Fig. 3.58 Relaxation curve superimposed on the load-strain curve of a geosynthetic.

Although creep and relaxation are different mechanisms in detail, they follow similar patterns. Materials which creep less also relax less.

To measure stress relaxation a specimen of material is stretched to the required strain as for a creep test, but then not allowed to move. This requires rigid clamp grips, a rigid framework and a very stiff load cell, and for this reason the measurement is not frequently performed.

It is simpler to estimate the stress relaxation from the creep isochronous lines. A vertical line drawn at the appropriate strain level will provide an estimate of stress relaxation that, if not perfect, is sufficient for design (figure 3.59) (Greenwood 1990, Walters et al. 2002). This can be used to determine, for example, whether the load in the geosynthetic remains above a given level throughout the design life. In figure 3.59 it is assumed that the strain in the geosynthetic at the end of construction, which is taken as the 100 h isochronous line, is 10% and the load correspondingly 19.5 kN/m. After a lifetime of 1000000 h, however, the remaining stress in the geosynthetic is predicted be 10 kN/m. Figure 3.60 depicts the reduction in prestress. This model is a simplified approximation: a more detailed analysis requires a more sophisticated model of material behaviour. This is explained in the following section.

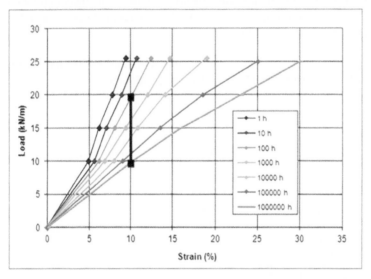

Fig. 3.59 Relaxation line on isochronous diagram. If the strain is 10% after 100 hours and no further movement is allowed, the load is predicted to fall along the vertical line from 19.5 kN/m to 10 kN/m after 1 000 000 h.

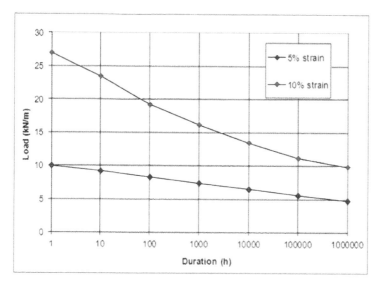

Fig. 3.60 Schematic example of relaxation in stress in the geosynthetic in figure 3.59.

## 3.4.5 *Modulus following sustained loading: seismic design*

The creep tests described, and the isochronous diagram that depicts the resulting strain, apply only to sustained loading and, as an approximation, to stress relaxation. They say nothing about the behaviour of the material afterwards.

So far we have considered creep strain under sustained load, creep-rupture and stress relaxation. We have also noted that the original strength is retained until shortly before its predicted creep-rupture lifetime (figure 3.51) (Greenwood 1997).

What happens, however, if a geosynthetic that has been under load for a period is then subject to a higher or a dynamic load? This is particularly important for seismic design, where the questions asked of a geosynthetic under sustained load are:
- How will the geosynthetic respond to the dynamic loading of an earthquake?
- How strong is the geosynthetic under rapid loading?

Understandably it is Japanese authors who have most interest in this and who have measured these properties in detail (Hirakawa et al. 2003, Kongkitkul et al 2004, 2007a, 2007b, 2010).

They found that:
- Provided that the geosynthetic is not close to creep-rupture, its strength is independent of its creep or dynamic loading history and depends solely on the rate of loading.
- The instantaneous stiffness bears no relation to the isochronous curves but is equal to, or even greater than, its initial value at the corresponding strain rate. This effect has also been observed by Voskamp (2001b, 2006) and is illustrated in figure 3.62 .

- If monotonic (ramp) loading is interrupted by a period of creep or dynamic loading and is then resumed, the strain will revert towards the original load-strain curve. Seen in this way, creep strain does not count as irreversible degradation but more as a reversible change of state.

Note also that the rate of loading of a geosynthetic during a seismic event is much higher than the rate of loading in conventional reference tensile test protocols. Consequently the stiffness of the geosynthetic is greater. This is more pronounced for drawn polyolefin geogrids than for polyester geogrids (Bathurst and Cai 1994).

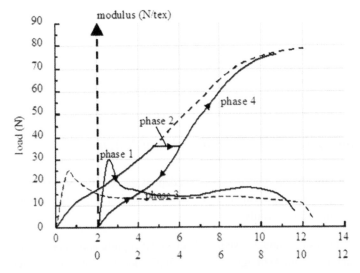

Fig. 3.61 Loading path during creep tests, with phase 2 as continuous loading path. Phase 1 and 2 cover the initial loading and creep phase. Phase 3 and 4 are the unloading and loading phase up to break. These last phases equal a stress strain curve with an origin at 2% (Voskamp and Van Vliet 2001a).

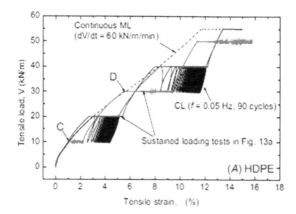

Fig. 3.62 Tensile-load strain relationship for HDPE geogrid (Kongkitkul et al 2004).

Figures. 3.61 and 3.62 show the stress-strain behavior of a geosynthetic subjected to a sequence of a) monotonic loading b) creep c) dynamic loading d) monotonic loading. Note that the stiffness (gradient) of the dynamic loading is greater than would be expected from the original stress-strain curve, and that in the final monotonic loading the curve approaches that which would have been reached had the original monotonic loading not been interrupted.

Hirakawa et al (2003) have set out a mathematical model to describe this behaviour. Following the methodology for soils rather than for polymers, the load is split into 'inviscid' and viscous components acting in parallel. The inviscid contribution to the load is proportional to the 'irreversible' strain $\varepsilon^{ir}$. For most polymers the viscous component is proportional to $\varepsilon^{ir}$ multiplied by a function of the rate at which it changes with time, $d\varepsilon^{ir}/dt$. The viscous behaviour is characterised by the proportional change in load following a change in strain rate, the constant of proportionality $\beta$ lying between 0.07 and 0.16 for a wide range of materials.

During monotonic loading both the inviscid and viscous components contribute. Under sustained load the response is viscous, while under dynamic loading the principal strain response is inviscid, with a smaller viscous element which causes a gradual increase in mean strain. The model has successfully described the behaviour of different polymer geosynthetics under complex loading patterns.

The authors have proposed an alternative method for predicting creep-rupture based on this model and considering simultaneous environmental degradation and sustained load, but assuming a constant strain at rupture. This method indicates that the simple reduction factor approach is conservative (Kongkitkul et al 2007b).

### 3.4.6   The effect of temperature on creep strain and creep-rupture

Conventional creep testing is performed at 20 °C, and therefore most conventional time-temperature shifting and SIM is referred to that temperature. The design temperature may however be different, for example 30 °C or even higher near the surface in tropical climates or 10 °C at depth in Northern Europe. The actual temperature will fluctuate, but since higher temperatures will accelerate creep to a greater extent than low temperatures will retard it, North American practice is to take the design temperature as halfway between the average yearly air temperature and the normal daily air temperature for the warmest month at the wall site (WSDOT 2005, AASHTO 2010). An advantage of time-temperature shifting and SIM is that creep strain can be predicted for different design temperatures by reading off the appropriate shift factor.

For example, figure 3.49 shows the thermal shift factor plotted against temperature. If the design temperature is 27 °C the thermal shift factor can be read off as equal to 0.7 which should then be deducted from the logarithms of the times on the master curve. The times themselves are then divided by 100.7 or 5.0; the predicted strain for 1 000 000 h at 20 °C becomes that for 500 000 h at 27 °C, i.e. creep proceeds faster at 27 °C. Extrapolation to 5 000 000 h would be necessary to predict the strain after 1 000 000 h at 27 °C. If the design temperature is 10 °C, figure 3.49 can be extended to give a shift factor of -1.0 and the times are then divided by $10^{-1}$ or multiplied by 10+1 = 10. In this case creep proceeds more slowly. The strain for 1 000 000 h at 20 °C becomes that for 10 000 000 h at 10 °C and the strain after 1 000 000 h at 10 °C is that predicted for 100 000 h at 20 °C.

Creep-rupture data can be shifted in a similar way and an example is given in section 3.7.2.2.

### 3.4.7   Effect of the environment and soil on creep strain

Creep strain is generally insensitive to limited weathering, chemical and biological effects. In addition, creep strains are in general not affected by installation damage, unless the damage is severe, or unless the load level applied is very near the creep limit of the undamaged material. In most cases, the load level applied is well below the creep limit of the material. See Allen and Bathurst (1996) for additional details on this issue. Thus, no further adjustment is generally required beyond the effect of temperature.

Note, however, that artificially contaminated soils may contain chemicals, such as organic fuels and solvents, which can affect the creep of geosynthetics, particularly polyolefins. If necessary, perform a short-term creep test on a sample of geosynthetic that is immersed in the chemical or has just been removed from it. If the creep strain is significantly different, do not use this geosynthetic in this soil.

Allen and Bathurst (Allen et al. 2003, Bathurst et al. 2005) collected data on reinforced walls and noted that reinforcement strains were generally less than 3%, that the rate of creep decreased with time, that there was no sign of failure in the backfill and that post-construction deformation was less than 30 mm in the first 10 000 h. Many other papers report that the strains measured are considerably less than those calculated in design, because at low strain less than about 3% part of the load is supported by the soil itself. They are also well below the level at which failure could take place. The effect of the soil is particularly marked for nonwovens. The soil is capable of taking a small tensile stress and much larger shear and compressive stresses, reducing the load on the geosynthetics. However, direct testing of creep in soil has proved unrepresentative and frequently influenced by the apparatus used. Creep measurements made in air are reproducible and representative but are necessarily conservative. Finally, where comparisons between in-soil and in-air creep measurements have been attempted there is evidence that the in-soil creep is never greater than in-air creep for the same tensile load and hence creep data used from in-air tests will always lead to safe design outcomes (Walters et al 2002).

### 3.4.8 Material effects in creep and relaxation

Creep in polymers is due to a slow rearrangement of the polymer chains, and at high loads to a successive breakage of bonds. Most polymers used in geosynthetics are semi-crystalline, and when loaded the crystalline areas remain largely unchanged.

Rearrangement occurs in the amorphous regions. At the glass transition temperature $\theta_g$ these amorphous regions change from a glassy to a rubbery state. Although the overall integrity of the polymer is maintained by the crystalline regions and the 'tie' molecules that join them across the amorphous regions, creep is more pronounced and more temperature-sensitive in polymers such as polyethylene and polypropylene which are used above $\theta_g$ than in those such as polyester which are used below $\theta_g$.

In polyester fibres ramp loading causes the ester groups in the polymer chain to change their local arrangement, resulting in a temporary reduction in stiffness and in the characteristic S-shaped (sigmoidal) stress-strain curve (see figure 3.29). This process of rearrangement continues even if ramp loading is interrupted, leading to a higher creep rate in this region followed by an increased stiffness as described in 3.4.5.

In the above methods rupture may occur preferentially at the nodes and joints within the structure of the geosynthetic, particularly in a geogrid. This is taken into account by specifying that a representative number lie within the gauge length used for measurement. The connections between the geosynthetic and a wall or other fixture are not considered, although it may be appropriate to add a reduction factor as in the German system (see chapter 3.7).

Extruded polyethylene geogrids are manufactured by stretching a perforated sheet. The material between the perforations is stretched into ribs in which the material is highly oriented and joined by thicker 'nodes' where there is only light orientation. The creep of these geogrids includes a continuation of this drawing process at the transition zone between node and rib, and rupture, if it occurs, typically takes place in this zone.

In the above creep and creep-rupture are considered as purely mechanical effects independent of the chemical degradation that may be taking place at the same time. No work has been carried out on the synergy of chemical and mechanical degradation. Only Zhurkov's formula (see section 3.4.3.5) assumes that breakage of chemical bonds may be a thermally activated effect accelerated by the presence of stress. This is a subject for future research.

**Summary**
- Measurements of tensile creep are required for many reinforcing applications.
- Creep strain refers to the elongation under load, the requirement being generally a maximum acceptable strain or a maximum strain following construction.
- Because of the long lifetimes creep data have to be derived by extrapolation or by performing accelerated tests. A typical set of tests would be at 10%, 20%, 30% and 40% of tensile strength lasting 1000 h for index testing and 10000 h for design. The latter should be supplemented by accelerated tests such as SIM.
- Creep in a reinforced structure can only occur if the geosynthetic is free to extend. If its dimensions are fixed it will undergo stress relaxation, or a reduction in prestress, which can be predicted from creep measurements. In many reinforcing applications the geosynthetic experiences a combination of both effects.
- Creep strain is sensitive to temperature but not to other environmental effects. Temperature provides the principal means of acceleration.
- At higher loads sustained loading can lead to creep-rupture, the consequence of which could be catastrophic. To predict the maximum load which will not lead to rupture during the design life of the geosynthetic it is necessary to perform at least twelve tests over a range of high loads to derive the relation between load and time to failure, and then extrapolate this to the design life. Accelerated tests can assist with this.
- The result is expressed as a reduction factor (RFCR) for a specific design life and temperature, which can then be combined with factors representing the effects of damage, weathering and the environment.
- Following a period of sustained or dynamic loading the stiffness of a geosynthetic can increase. The residual strength does not change until shortly before the creep-rupture lifetime. These effects are important for seismic design.

# References

*Chapter 3.4*
- AASHTO: American Association of State Highway and Transportation Officials. 2010. LRFD Bridge Design Specifications. 5th edition, Washington, DC, USA.
- Allen T M, Bathurst R J. Investigation of the combined allowable strength reduction factor for geosynthetic creep and installation damage. Geosynthetics International 1996 Vol 3 No 3, pp 407-439.
- Allen, T.M. and Bathurst, R.J., 2002, "Observed Long-Term Performance of Geosynthetic Walls and Implications for Design", Geosynthetics International, Vol. 9, Nos. 5-6, pp. 567-606.

- Allen T M, Bathurst R J, Holtz R D, Walters D, Lei W F. A new working stress method for prediction of reinforcement loads in geosynthetic walls. Canadian Geotechnical Journal, Vol 40, 2003, pp 976-994.
- Bathurst, R.J., and Cai, Z. 1994. In-isolation Cyclic Load-Extension Behavior of Two Geogrids, Geosynthetics International, Vol. 1, No.1, pp. 3-17, 1994.
- Bathurst R J, Allen T M, Walters D L. Reinforcement loads in geosynthetic walls and the case for a new working design stress method. Geotextiles and geomembranes, Vol. 23, 2005, pp287-322.
- Bathurst, R.J., Vlachopoulos, N., Walters, D.L., Burgess, P.G. and Allen, T.M. 2006. The influence of facing rigidity on the performance of two geosynthetic reinforced soil retaining walls, Canadian Geotechnical Journal, Vol. 43, No. 12, pp. 1225-1237.
- Bathurst, R.J., Huang, B. and Allen, T.M. 2012. Interpretation of laboratory creep testing for reliability-based analysis and load and resistance factor design (LRFD) calibration, Geosynthetics International, 19(1) (in press).
- Greenwood J H, The creep of geotextiles, 4th International Conference on Geotextiles, Geomembranes and Related Products, The Hague, 1990, Balkema, Rotterdam, 1990, pp645-650.
- Greenwood, J.H., 1997, "Designing to Residual Strength of Geosynthetics Instead of Stress-Rupture", Geosynthetics International, Vol. 4, No. 1, pp. 1-10.
- Greenwood J H, Kempton G T, Watts G R A, Bush D I, Twelve year creep tests on geo-synthetics reinforcements, Proceedings of the Second European Geosynthetics Conference, Bologna, 2000, pp 333-336.
- Greenwood, J.H., Kempton, G .T., Watts, G.R.A. and Brady, K.C. 2004. Comparison between stepped isothermal method and long-term tests on geosynthetics. 3rd European Conference on Geosynthetics, Munich, Germany, 2004, ed. Floss, R., Bräu, G., Nussbaumer, M., Laackmann, K. pp 527-532.
- Hirakawa D, Kongkitkul W, Tatsuoka F, Uchimura T. Time-dependent stress-strain behaviour due to viscous properties of geogrid reinforcement. Geosynthetics International, Vol 10, 2003, pp 176-199.
- Kongkitkul W, Hirakawa D. Tatsuoka, F, Uchimura T. Viscous deformation of geogrid reinforcement under cyclic loading conditions and its model simulation. Geosynthetics international, Vol. 11, 2004, pp 73-99.
- Kongkitkul W, Tatsuoka F. A theoretical framework to analyse the behaviour of polymer geosynthetic reinforcement in temperature accelerated creep tests. Geosynthetics international, Vol. 14, 2007a, pp 23-38.
- Kongkitkul W, Tatsuoka F, Hirakawa D. Creep rupture curve for simultaneous deformation and degradation of geosynthetic reinforcement. Geosynthetics International, Vol. 14, 2007b, pp189-200.
- Kongkitkul, W., Tatsuoka, F., Hirakawa, D., Sugimoto, T., Kawahata, S. and Ito, M. 2010. Time histories of tensile force in geogrid arranged in two full-scale high walls. Geosynthetics International, 17(1): 12–33.

- Koo H J, Cho H W, Choi S W, Kim S H. Comparison of various creep aspects for determining creep reduction factors of geogrids. The First Pan American Geosynthetics Conference & Exhibition 2-5 March 2008, Cancun, Mexico, pp 591-597.
- Lothspeich S E, Thornton J S. Comparison of different long term reduction factors for geosynthetic reinforcing materials. Proceedings of the Second European Geosynthetics Conference, Bologna, 2000, pp 341-346.
- McGown, A., Khan, A. J. & Kupec, J. (2004a). The isochronous strain energy approach applied to the load–strain–time–temperature behaviour of geosynthetics. Geosynthetics International, 11, No. 2,114–130.
- McGown, A., Khan, A. J. & Kupec, J. (2004b). Determining the design parameters for geosynthetic reinforcements subject to single-stage actions using the isochronous strain energy approach. Geosynthetics International, 11, No. 5, 355–368.
- McGown, A., Kupec, J. & Khan, A. J. (2004c). Determining the design parameters for geosynthetic reinforcements subject to multi-stage actions using the isochronous strain energy approach. Geosynthetics International, 11, No. 6, 455–469.
- Voskamp W., van Vliet F., Variations of Creep Rate at constant Loading of PET Geogrid Strapping, 3rd International Symposium on Earth Reinforcement, Kyushu, 2001a, pp 159-165.
- Voskamp W, van Vliet F., Retzlaff J. Residual strength of PET after more than 12 years creep loading, 3rd International Symposium on Earth Reinforcement, Kyushu, 2001b, pp 165-170.
- Voskamp W., Retzlaff J., Swing up in creep curves at high loadings of PET bars – further proof of the molecular chain change model, Proceedings of the 8th International Conference on Geosynthetics , Yokohama, 2006.
- Walters, D.L., Allen, T.M., and Bathurst, R.J. "Conversion of Geosynthetic Strain to Load using Reinforcement Stiffness", Geosynthetics International, 2002, Vol. 9, Nos. 5-6, pp. 483-523.
- Wrigley N E, Zheng H. A new insight into the creep of HDPE geogrids. The First Pan-American Geosynthetics Conference and Exhibition, Cancun, Mexico, 2008, pp 617-623.
- WSDOT: Washington State Department of Transportation, 2005. Standard Practice T925. Standard practice for determination of long-term strength for geosynthetic reinforcement. Revised 11-23-05.
- Yeo S-S, Hsuan Y.G. Evaluation of stepped isothermal method using two types of geogrids 2008, EuroGeo4, Edinburgh, UK, paper 285.
- Zhurkov S. Kinetic concept of the strength of solids. International Journal of Fracture Mechanics, 1965, Vol 1, pp311-323.

# 3.5 Compressive creep

J.H. Greenwood

## 3.5.1 Drainage geosynthetics

As introduced in chapter 2.1, the purpose of drainage geosynthetics is to allow water or, in the case of landfill, leachate, to separate from the soil and to flow through the geosynthetic into a pipe or channel at its edge. They provide an economical replacement for the traditional layers of coarse gravel. Examples are shown in figures 3.63 and 3.64.

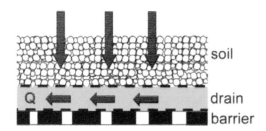

Fig. 3.63 Example of a drainage composite laid horizontally.

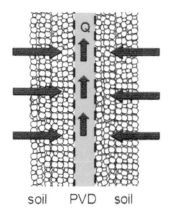

Fig. 3.64 Example of a drainage composite laid vertically, in this figure a Prefabricated Vertical Drain (PVD) which transports the water from the soil to the surface. In a vertical drainage mat, e.g. behind a wall the water flows to the bottomdrainage pipe.

For two-sided drainage the geosynthetic will consist of two sheets of nonwoven fabric separated by an openwork core (Geospacer), to which the nonwoven sheets are heat bonded or sewn. For one-sided or basal drainage it will consist of one sheet of permeable nonwoven and one sheet of impermeable geomembrane separated in the same way. The nonwoven sheets act as filters, stopping soil particles from entering the core, while the core allows the remaining liquid drain laterally. It can also increase the friction between itself and a geomembrane. All kinds of shapes and materials are used to form the geospacer: for example networks of stiff randomly oriented fibres, geonets and corrugated or embossed thermoplastic sheets.

Some drainage materials are intended only to drain the soil following installation and are then redundant. These applications do not concern us. In civil engineering structures such as road embankments the planned lifetime of the structure can be as much as one hundred years. In landfill applications drainage geosynthetics are used both at the bottom, to drain leachate emanating from the waste material, and at the top, to divert and collect surface water. These functions continue throughout the design life of the landfill, which may be planned at several hundred years.

The hydraulic design of these drainage layers, necessary to ensure that the layer has sufficient capacity to provide drainage at each location across its area, is described in detail by Giroud et al. (2000a). With regard to long-term durability it is obvious that the geosynthetic must retain its flow capacity, which can be compromised by the following factors:

- Instantaneous compression of the core.
- Instantaneous intrusion of the filter into the core.
- Time-dependent compression of the core (creep).
- Time-dependent intrusion of the filter into the core.
- Chemical degradation of the core.
- Clogging due to fine soil particles that penetrate the filter.
- Clogging by biological growth or root penetration.

We must assume that the first two instantaneous factors have been taken into account in the initial design, given that the initial testing has been performed under suitably representative conditions. Giroud et al. mention a 'seating time'. In this chapter we shall consider time-dependent creep and intrusion. These, together with chemical degradation and clogging, lead to a series of reduction factors which will be described in chapter 3.7.

The slow compression of the drainage geosynthetic due to the pressure of the soil can lead to:
- Reduction in thickness, leading to reduction in the rate of flow.
- Intrusion of the filter with similar results.
- Collapse of the core which will prevent flow dramatically.

These are described further in the following sections.

### 3.5.2   *Measurement of compressive creep strain*

The slow reduction in thickness can be measured in a similar way to tensile creep and extrapolated to longer durations. A minimum acceptable thickness will be set based on the requirements of flow.

Time-dependent intrusion is more difficult to quantify. To simulate the behaviour in soft soils one can place the geosynthetic between sheets of foam which will intrude into the open spaces in the core. This again has to be related to the reduction in flow capacity and thence to design life.

In the simple compression test creep strain is measured by placing a square of the drainage material between two flat plates, applying a controlled pressure and measuring the change in the separation of the plates by means of sensitive extensometers attached to them. Details are given in ISO 25619-1 (formerly EN 1897). At small loads creep in compression is similar to that in tension: there is an initial elastic reduction in thickness followed by a further reduction with time. The rate of creep reduces with time and is generally plotted as percentage compression (with zero at the top of the diagram) or relative thickness (100% minus percentage compression) against log (time) (figure 3.65). Different geospacers have different mechanical, flow and creep characteristics which should be taken into account during selection.

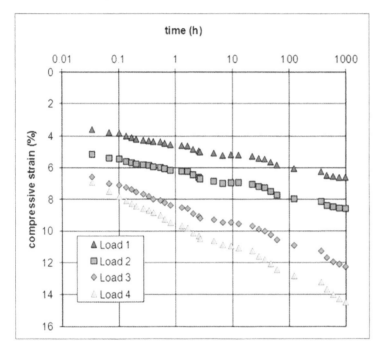

Fig. 3.65 Creep of a drainage composite at four different loads.

Straight lines or curves can be fitted to diagrams such as those in Fig 3.65. In theory it would be more appropriate to fit a function which approximates to zero at infinite time, unlike a straight line which will cross the axis and predict negative strains. Müller et al (2008) predicted an increase in gradient of $d\varepsilon/d$ (log $t$) at longer times for some materials. In reality the acceptable thickness is unlikely to be less than half the original thickness, so that such refinements can be neglected.

Fig. 3.66 Testing compressive creep with the aid of SIM (courtesy ERA Technology Ltd).

Temperature can be used to accelerate compressive creep in the same way as tensile creep, including the stepped isothermal method (SIM) which has been standardized as ASTM D 7361 (Narejo and Allen 2004) (figure 3.66). SIM results can therefore also be used to predict the compressive creep performance at temperatures other than room temperature.

Greenwood and Young (2008) discuss the application of SIM in compression and note the following:
- The small thickness requires sensitive extensometry attached directly to the loading plates. Specimens can be stacked on top of each other to increase the thickness provided that they do not project into one another.
- Elimination of physical ageing requires that the material, particularly the core, should be more than one month old.
- A polyethylene core limits the maximum temperature to around 60 °C.
- The thermal mass of the compression plates substantially increases the time taken for the system to stabilize following a temperature step. Fins, similar to the cooling fins used in car engines and electrical motors, can be used to improve the heat transfer to the platens. Even better, heated platens ensure a rapid change in temperature across the whole 100 mm square cross-section.
- Some practice is required to optimize the procedure. The choice of shift factor can vary from one operator to the next: using the same procedure but different people to evaluate the results there was a standard deviation of 0.4% strain (95% confidence limit 0.8% strain).

Compressive creep also occurs in expanded polystyrene. While not directly related to the life prediction of geosynthetics, further information is given by Yeo and Hsuan (2006, 2008).

### 3.5.3 Relation between compression thickness and flow

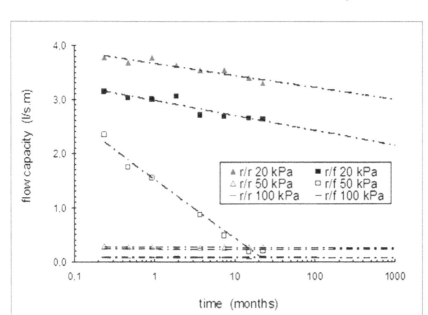

Fig. 3.67 Example of relation between flow rate and compressive stress. With r=rigid bedding, f=soft bedding (Böttcher 2006).

A typical drainage geosynthetic will be designed to collect liquid over a certain area and then to release it at one edge into a linear pipe or drain. The depth of the liquid will therefore increase toward the drainage edge at which point it reaches a maximum and then falls off rapidly. If the thickness of the geosynthetic is less than the maximum depth of the liquid it will constrict the flow. The minimum acceptable thickness could therefore be set equal to this maximum. For a detailed calculation of the flow characteristics see Giroud et al. (2000b).

One procedure is therefore to predict the thickness at the end of the design life by extrapolation of the compressive strain, and then to measure the flow capacity of specimens compressed to the same thickness (Müller et al 2008). The effect of soft bedding can be added by using appropriate plates when measuring flow capacity or, when this is not appropriate, by multiplying the ratio of initial flow capacities measured using the standard method with and without soft bedding. The change in flow capacity can also be expressed as a reduction factor in the same manner as for reinforcements in tension, and can then be combined with other ageing effects.
Böttcher (2006) performed long-term compressive tests using soft bedding and removed the samples from time to time to measure the changes in flow capacity, as shown in figure 3.67 soft bedding can have a critical effect on reducing the flow capacity.

### 3.5.4 Predicting collapse

Fig. 3.68 Collapse of a core. Top: before start of the test. Bottom, result after a compression of 420 kPa, core partly collapsed, discharge capacity still present (courtesy Colbond bv.).

Some types of core can collapse or become unstable after a certain length of time under high compressive loads (figure 3.68). The load-deformation-curve of a short-term test according to EN 25619-2 shows the probability of collapse. The time to collapse at longer durations or, conversely, the highest load that would not lead to collapse over the design life, can be predicted in the same way as predicting time to failure of a reinforcement in tension (creep-rupture) in chapter 3.4. The equipment is similar to that in ISO 25619-1. The time to collapse is measured at high loads, if necessary accelerated by using SIM, and a relation established between applied load and time to collapse. This can be extrapolated to give the maximum compressive load for a given design life.

This tendency to collapse depends on the physical structure of the core. One can also assume that collapse will commence, if only slowly, once the core has reached a certain critical thickness: the advantage of this approach is that this can be predicted more easily from creep strain curves without having to perform accelerated collapse tests. This approach disregards the additional effects of ageing and temperature, both of which must be considered in long-term prediction (Zanzinger 2008).

Since collapse will cause dramatic reduction of the flow its effects cannot be taken into account by a reduction factor applied to the flow capacity. In design the material must be selected to withstand the anticipated pressure of the soil over the entire design life, with an appropriate safety factor to allow for the uncertainty in prediction.

## 3.5.5   The effect of shear

Where the soil pressure is hydrostatic or perpendicular to plane of the drainage geosynthetic there is no shear force on it. If not, a shear component is added to the compressive force, a particular case being the sloping side of a landfill where the vertical pressure divides into compression proper perpendicular to the sloping surface and a shear force parallel to the slope. Addition of shear to the compressive load can cause a change in the rate of compressive creep, but its principal effect is on creep collapse.

Shear forces can distort or roll the core or can lead to separation of core from nonwoven. Some cores are more sensitive than others. The geosynthetic should be tested firstly in simple compression and then in compression with the addition of shear.

Two methods are used to measure the effect of shear:
- Application of a lateral force to the upper compressive plate, with freedom for the plate to move sideways. The ratio of shear to compressive load can be adjusted by choosing the lateral force.
- Compression between two inclined wedges, the tangent of whose angle gives the ratio of shear to compressive load. The sample must not be allowed to slip, either by ensuring sufficient friction or by a mechanical bond. The ratio of shear to compressive load can be adjusted by choosing wedges of different angles.

These methods are illustrated in figures 3.69 and 3.70.

Fig. 3.69 Equipment for measuring the combined effect of shear and compression on a drainage geosynthetic by adding a lateral force.

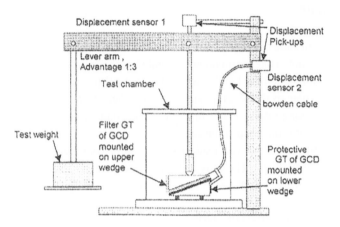

Fig. 3.70 Schematic diagram of equipment for measuring the combined effect of shear and compression on a drainage geosynthetic by using inclined wedges, Müller 2006.

Multiple tests must be performed to establish the relation between compressive load, shear load and time to collapse. This relation can then be extrapolated to the design life.

Müller et al. (2008) provide a comprehensive description of measurements of compressive stress, flow capacity, collapse and the effects of shear and temperature. They used wedges to give shear: compressive ratios of 1:2.5 and 1:3, with friction plates on both wedges, and immersion in water up to 80 °C. Their materials had a nonwoven on the upper surface and a geomembrane on the lower surface. They measured the vertical displacement of the upper surface, which also caused the area to become less. They noted that different styles of core, notably cuspated ones, differ greatly in their sensitivity to shear forces, but concluded that the cores they studied would only collapse under very high compressive and shear stresses at elevated temperatures.

**Summary**

- The lifetime of drainage composites can be limited by compression of the core, intrusion of the filter into the core, chemical degradation and clogging.

- Compression occurs instantaneously on loading and proceeds further with time, known as compressive creep. It may be necessary to measure separately how this will restrict flow. Compressive creep can be predicted by extrapolation of creep tests, accelerated if necessary by temperature. This can be used to predict the residual thickness at the end of the design life or, conversely, the maximum permissible soil pressure if the thickness of the geosynthetic is to remain above a minimum value dictated by the requirements of flow.

- It is important to include the effect of soft soils in any prediction. This can be simulated during measurement by compressing between layers of soft foam.

- Compressive loading can cause the geosynthetic to collapse and stop the flow completely. By performing multiple tests a relation can be established between compressive load and time to collapse, or alternatively the time to a given compressive strain, which can then be extrapolated to the design life to give minimum load that could cause collapse during that time.

- Shear loads, such as occur on the sloping sides of landfills, can accelerate collapse in certain designs of core. If necessary shear can be added to the compressive load when measuring the time to collapse.

# References

*Chapter 3.5*

- Böttcher R D. The long-term flow capacity of geocomposites. Proceedings of the 8th International Conference on Geosynthetics, Yokohama, Japan, ed. Kuwano J and Koseki J. 2006. Millpress, Rotterdam, pp 423-426.

- Giroud J P, Zornberg J G, Zhao A. Hydraulic design of geosynthetic and granular liquid collection layers. Geosynthetics International, 2000a, Vol. 7, pp 285-380.

- Giroud J P, Zhao A, Richardson G N. Effect of thickness reduction on geosynthetic hydraulic transmissivity. Geosynthetics International, 2000b, Vol. 7, pp 433-452.

- Greenwood J H, Young D K. Measuring creep of drainage materials by means of the stepped isothermal method in compression. EuroGeo4, Edinburgh, UK, paper 221 (2008).

- Müller W, Geosynthetic drain elements for landfill cap pings: Long-term water flow capacity and long-term shear strength, Proceedings of the 8th International Conference on Geosynthetics, Yokohama, Japan, ed. Kuwano J and Koseki J. 2006. Millpress, Rotterdam, pp 321-324.

- Müller W, Jakob I, Tatzky-Gerth R. Long-term water flow capacity of geosynthetic drains and structural stability of their drain cores. Geosynthetics International, Vol. 15, Issue 6, 2008, pp 437-451.

- Narejo D, Allen S. Using the stepped isothermal method of geonet creep evaluation. 3rd European Conference on Geosynthetics, Munich, Germany, 2004, ed. Floss, R., Bräu, G., Nussbaumer, M., Laackmann, K. pp 539-544.
- Yeo S-S, Hsuan Y G. The compressive creep behavior of an expanded polystyrene geofoam Proceedings of the 8th International Conference on Geosynthetics, Yokohama, Japan, ed. Kuwano J. and Koseki J. 2006. Millpress, Rotterdam, pp 1639-1642.
- Yeo S-S, Hsuan Y G. Evaluation of creep behavior of expanded polystyrenegeofoam using accelerated and conventional test methods. The first Pan-American Geosynthetics Conference and Exhibition, Cancun, Mexico, 2008, pp 633-641.
- Zanzinger H. Evaluation of the long-term functioning of drainage geocomposites on landfill cover systems. The first Pan-American Geosynthetics Conference and Exhibition, Cancun, Mexico, 2008, pp 585-590.

# 3.6    Mechanical damage, dynamic loading and abrasion

J.H. Greenwood

## 3.6.1    Prevention or measurement?

Damage during the transport and installation of a geosynthetic, particularly with coarse backfills, is undesirable but sometimes inevitable. Stones and harsh treatment result in visible cuts and holes, poor laying can leave behind folds and local stresses, while weld lines in geomembrane sheets change the structure of the polymer itself. While these are not directly issues of long-term degradation, these sites can and will lead to early failure. In fact, exhumation after a number of years often shows that mechanical damage is the only form of degradation that can be identified with certainty.

This damage can change the hydraulic properties, and can lead to a general reduction in mechanical strength. It also provides sites, not just holes but stressed areas or wrinkles, from which other forms of degradation can start, such as chemical attack and, in geomembranes, environmental stress cracking. Should we do all we can to prevent damage, or should we accept its existence and make allowances for it? And how does damage affect lifetime?

This depends on the application. Holes and cuts will reduce the effectiveness of filters and separators. Reinforcements will lose strength. In all applications damage can be reduced, if not avoided, by careful selection of the material and control of the method of installation. In general the effect of damage cannot be sufficiently quantified for it to be included in a prediction of lifetime - more often it is revealed as the cause of unexpected premature failure. In the case of reinforcing geosynthetics, however, it may be possible to quantify the reduction in strength by measuring it. This assumes that the mechanical loads in service are much less than those during the short period of installation, and that the damage they cause is insignificant. Where, as under railways and in some road applications, the continuous dynamic loading can generate further damage, the lifetime will be governed by abrasion and fatigue.

## 3.6.2    Nature of damage

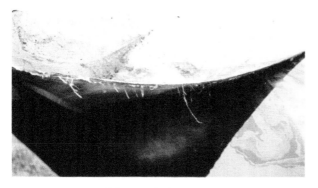

Fig. 3.71 Example of damage during installation.

255

Visible damage consists of the following:

- Small holes are produced in sheet materials as well as in geomembranes. They will change the hydraulic properties, reduce the strength and, in some cases, reduce the stiffness.
- Local abrasion and scuffing will occur. This could change the hydraulic properties or reduce the strength.
- Splitting can occur in split film woven materials (figure 3.72).
- Cuts in fibres, yarns or the ribs of geogrids will reduce the strength and may initiate further tearing.
- Cutting or removal of coatings and protective sheaths, particularly in geogrids, expose the central fibres to weathering and to chemical attack.

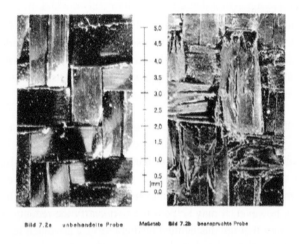

Bild 7.2a    unbehandelte Probe        Maßstab    Bild 7.2b    beanspruchte Probe

Fig. 3.72 Mechanical damage on a woven split film polypropylene (courtesy J. Müller-Rochholz). Left: virgin specimen. Right: damaged specimen.

At the microscopic scale fine fibres can be damaged by small abrasive particles that penetrate between them, an effect also found in industrial filters (Ehrler and Gündisch 1999). Fine fibres can be given an individual coating, while in strips and geogrids a bundle of fibres may be protected by a coating or sheath. The thicker the coating, the greater the protection.

### 3.6.3  Robustness classes

Intuitively, thicker geosynthetics are less susceptible to damage. Needlepunched fabrics are generally thicker and therefore more damage resistant than thinner heatbonded nonwovens. Equally obviously, larger and sharper stones in the backfill cause more damage, such as the when backfill is dropped from a height on to a geosynthetic, even more when large rocks or cement blocks are placed on a geosynthetic for erosion control or when vehicles are driven over the geosynthetic before construction is complete. Conversely clays and sands, with median particle diameters no greater than 2 mm, generally produce little mechanical damage.

There are therefore classes of robustness in which the strength and the mass per unit area of the geosynthetic is matched to the severity of the application. Based on original tests (Koerner and Koerner 1990), it is assumed that is a geosynthetic of the correct class is selected it will prove sufficiently resistant to damage during construction and in the long term. These classes, however, do not apply to reinforcements. In Germany a classification system is used for the selection of geosynthetics in road building (FGSV Merkblatt 2005). The 5 classes of geosynthetics are defined with separate requirements for non wovens, PP and PE split fiber fabrics and PET multifilament fibers. The required class depends on the fill material on top of the geosynthetic and the installation method.

### 3.6.4   *Simulated test to measure the effect of damage in reinforcements*

The reduction in strength of a reinforcement is measured by direct, if simulated, installation. A layer of backfill, which can be a well sieved and standardized aggregate or material taken directly from a site, is placed and compacted on a hard surface (figure 3.73). Samples are placed on it, following which a second layer of material taken from site (unless another backfill is specified) is laid over the samples and compacted in the same way (figures. 3.74, 3.75). The samples are then excavated carefully to avoid any further damage, examined and then tested for loss of strength (figure 3.76).

The procedure is described in BS 8006 Annex D and in ASTM D 5818. General guidance is given in ISO 13437 and in the FGSV Merkblatt (2005). The number, size and orientation of the samples should be chosen to suit the subsequent tensile tests, while samples of undamaged material taken from the same roll should be retained for comparison. Since the damage is random and sometimes widely spaced, the samples should be no narrower than 200 mm and it may be necessary to test more than the traditional five specimens in order to obtain a representative mean strength for the damaged material and a statistically significant comparison.

Fig. 3.73 Smoothing the first layer of backfill (ERA Technology Ltd/Geoblock Ltd).

Fig. 3.74 Placing the aggregate on the samples (ERA Technology Ltd/Geoblock Ltd).

Fig. 3.75 Compacting the second layer of backfill (ERA Technology Ltd/Geoblock Ltd).

Fig. 3.76 Final excavation using hand tools (ERA Technology Ltd/Geoblock Ltd).

Record the nature and thickness of both layers of backfill, the mass of the compacting equipment and the number of passes of that equipment. Excavate the samples with great care, removing most of the backfill with mechanical equipment but finish the excavation with hand tools. Do not be tempted to tug at a corner of the sample to free it from the backfill. It is critical not to damage the sample further.

Having washed away the remaining soil particles, the population and size of holes in a damaged geosynthetic can be evaluated visually or by quantitative measurement. For reinforcements measure the tensile strength of both the damaged and the undamaged samples. The retained strength is the ratio of the strength of the damaged material to that of the undamaged, i.e. less than unity, while the reduction factor $RF_{ID}$ is defined as the strength of the undamaged material divided by the strength of the damaged material and is thus greater than unity. The modulus of the geosynthetic should not be affected by small amounts of damage, although there may be other effects such as the accumulation of small fines in the open spaces of a nonwoven or the flattening of a woven structure. For non-reinforcing applications, particularly with nonwovens, measure burst strength. Permeability or another hydraulic property could be measured if appropriate.

Record the following:
- the nature of the backfill both below and above the sample: particle size distribution, hardness and angularity;
- the depth at which the sample is installed;
- whether the material is driven over by vehicles before compaction;
- method and degree of compaction;
- The mean strength (or other property) of undamaged and damaged samples, and the ratio of undamaged to damaged, $RF_{ID}$.

### 3.6.5  Laboratory damage test

The laboratory damage test EN 10722 is similar to the simulated site test but performed on a smaller scale in the laboratory. An open square frame is filled with aluminium oxide aggregate of a defined size and standardized hardness which is then compacted. A sample of suitable dimensions for tensile testing is laid across the frame. A second open frame is laid over the specimen and filled with aggregate (figure 3.77). A plated of set dimensions is then used to compact the aggregate to a specified load and number of cycles. The aggregate is then removed and the specimen tested for tensile strength or another property if appropriate. RFID is deduced in 3.7.2.3.

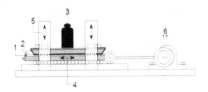

Fig. 3.77 Laboratory damage test.

EN 10722 has proved useful for comparing materials of similar nature (e.g. different woven polypropylens). Graphs showing the level of correlation and scatter between laboratory and site damage tests, using site materials instead the aluminium oxide, are given by Hufenus et al (2005). Wovens and nonwovens show more damage in the laboratory test, due probably to damage by sharp particles that infiltrate the fine structure. Comparison with crushed stone was satisfactory, with the exception of a yarn reinforced nonwoven. The authors conclude that the method provides a cheap means of predicting site damage and determining $RF_{ID}$.

The results of EN 10722 and therefore any indication on a specification sheet based on EN 10722 are not generally accepted for design unless the correlation has been made for the particular product and the particular type of fill.

### 3.6.6   *Effect of damage on long-term strength*

Damage is regarded as being not dependent on time. One might intuitively expect damaged geosynthetics to rupture relatively sooner than undamaged ones, as the damage would initiate tearing. This can occur in geomembranes and other plastics with no significant orientation such as gas pipes. Much work has been performed to test this hypothesis, but currently there is no evidence of any synergy. The creep-rupture behaviour of moderately damaged material is no worse than is predicted by multiplying the two reduction factors $RF_{CR} \times RF_{ID}$, though for heavily damaged material the lifetimes may be shorter than predicted by this method This has been validated by a number of accelerated tests (Greenwood 2002, Paula et al. 2008).

In geomembranes it is necessary to test for environmental stress cracking (ESC), when a notch caused by damage or partial welding is stressed. This was the cause of early failures of geomembranes and was detected at the nodes of some early PE geogrids and is the most common cause of premature failure in plastics in general. Tests for this have been standardized as EN 14576: 2005 and ASTM D 5397. Geotextiles and geogrids consist of oriented material and are generally insensitive to ESC.

## 3.6.7   Dynamic Loading

In addition to static loads due to the soil or to prestress during construction, geosynthetics in rail and road applications will experience dynamic loading due to the traffic above them. The loading will be in compression, although for a reinforced soil structure some of this will be transferred to tension in the reinforcements. Damping and hysteresis will reduce the sharpness of the loading. As a result, while the static loading increases with depth the dynamic loading may decrease, as shown in figure 3.78.

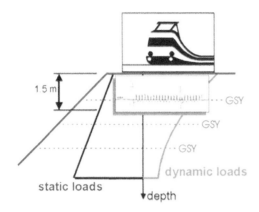

Fig. 3.78 Schematic diagram of dynamic and static loading in a reinforced soil structure (courtesy H. Zanzinger, SKZ).

The effect of the compressive dynamic loading will be to induce mechanical damage in the same manner as for static loading. This damage can impair the function of the geosynthetic as a filter or separator, and it can reduce the tensile strength of a reinforcement. A typical example would be a separator under railway ballast, whose function is to prevent fines from the soil penetrating the ballast. This is generally avoided by choice of a suitably robust material or by providing a protective layer in addition to the separating layer.

The effect of the tensile dynamic loading, superimposed on the existing static loading, can be to reduce the strength of the reinforcement to the point where it fails, although in reinforced soil the effect of static loading generally dominates. Degradation due to dynamic loading is also known as fatigue, a property that has been very widely investigated in engineering materials. Testing consists almost exclusively of the application of high dynamic loads, measurement of the number of cycles to failure, establishment of a relation between load and the number of cycles, and the extrapolation of this to the lower service loads in order to predict the number of cycles to failure. In some materials there may be a fatigue limit, a load below which there is effectively no degradation. The dynamic loads may be purely tensile, tensile and compressive about zero, or may include a static component so that they oscillate between an upper and lower load. The relation between load and cycles is referred to in Germany as a Wöhler curve after the 19th century German engineer August Wöhler.

The frequency of the dynamic loading gives an opportunity for accelerated testing. While the frequency of the dynamic loads due to traffic may be less than 1 Hz, dynamic testing can be performed at higher frequencies provided that it can be shown that the higher strain rates do not change the performance of the materials and that the energy dissipated does not lead to significant heating.

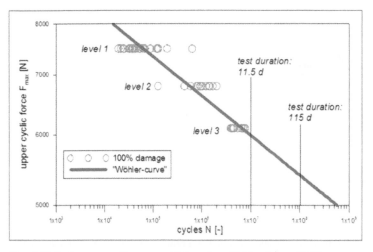

Fig. 3.79 Fatigue curve for a geogrid (courtesy H. Zanzinger, SKZ).

The lifetime of reinforcing geosynthetics loaded in tension can be predicted in this way. Figure 3.79 shows the fatigue curve for a geogrid. Multiple tests were performed at three different upper loads, with the lower load equal to half the upper load. This is denoted by the ratio R = 0.5. The frequency used was 10 Hz. The number of cycles to failure (or 100% damage) is measured and plotted logarithmically. The result is a straight line which can then be extrapolated down to the service load.

Measurement of the properties of the geosynthetic during fatigue testing such as amplitude, energy expended and temperature can provide an early warning of the onset of damage. For a polyester geogrid Zanzinger (2007) identified a unique point at which all three properties were at a minimum for each loading condition and defined this as 'the onset of damage'. This leads to a second fatigue curve, parallel to the first, and at about 20% of the lifetime, with this as the end point instead of ultimate failure (figure 3.80). Similarly, Retzlaff et al. (2008) noted changes in the modulus of polyester during dynamic loading, as occurs after sustained loading. They also used infra-red and thermal techniques to detect changes in the microstructure of polyolefins. These provide alternative end points and early warning criteria for fatigue testing.

Ageing and oxidation have to be considered separately.

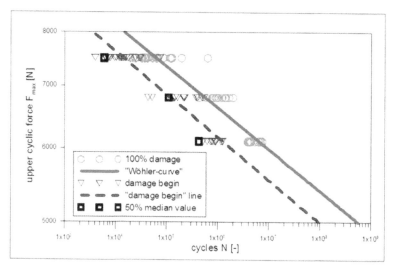

Fig. 3.80 Fatigue curve for a geogrid showing the fatigue curve for the beginning of damage (cortesy H. Zanzinger, SKZ).

## 3.6.8 *Abrasion*

Abrasion degrades geosynthetics in applications such as coastal erosion where the geosynthetic experiences particles in motion. EN 13427 provides an index test for resistance to abrasion using a sliding block test. The same method could in principle be used to establish a relation between, for example, applied load and time (or cycles) to failure. This could then be extrapolated to a given design life in order to determine the maximum load that could be applied in service.

At present durability is established by selecting appropriate materials that satisfy EN 13427 rather than by attempting to determine design life.

## 3.6.9 *Protection efficiency*

In many applications geomembranes, or drainage layers above geomembranes, have to be protected against damage, whether due to sharp stones in a fill or to solid waste in a landfill. This is achieved typically by a layer of sand, a thick nonwoven geotextile or a combination of both. A geogrid sandwiched between two lighter nonwovens could be used as well. The technology and properties of protective layers is excellently described by Müller (2001). This discussion will be limited to aspects of life prediction.

In the simplest case the geotextile simply has to protect the geomembrane against perforation, such that it can withstand the local tensile stresses and stop them being transferred to the membrane. For long-term durability, however, geomembranes have to be protected from excessive local strains. Under sustained tension a thermoplastic can develop cracks earlier and at lower strains than would be predicted from simple tensile or short-term creep tests. This 'brittle' failure or stress cracking has been investigated in detail and, as a general rule, the strains in a geomembrane should be kept below 3-5%. In the presence of certain liquids, mostly organic liquids or surfactants, cracking can occur at even lower strains. These effects are suppressed in highly oriented plastics such as geotextiles.

Local point loads will be caused by sharp stones or solid objects in a landfill. ASTM D5514 provides a test method for resistance to perforation. A thick protective geotextile will spread this compressive load over a wider area. Nevertheless, this compressive load can lead to an indentation in the geomembrane and this indentation will result in a biaxial tensile stress in the geomembrane at its centre, while at its edges of the indentation, where the geomembrane is in flexure, the upper surface of a thick membrane will be in tension. These local strains can easily exceed 3%. This has led to the European test method EN 13517, in which a protective geotextile is placed over a geomembrane under which a soft metal plate records the depths of the indentation. 20 mm steel balls are used to simulate the fill and the support layer is a rubber pad, unless site-specific materials are required. The test is performed over 12 h at 20 °C at a load of 300 kN/m². The deformation of the soft metal plate is determined to establish the maximum local elongations and radii of curvature. For an accelerated long-term test Brummermann (1997) conducted the test over 1000 h at 40 °C at 1.5 × the anticipated service load, while a more stylised machined indentation pad can also be used. A maximum deformation of 0.25% is recommended, leading to maximum edge fibre strains of 3-5%.

These procedures do not lead to numerical life prediction. They do not consider the effect of ageing and oxidation. Instead, they assume a maximum long-term strain for the geomembrane and then assure that the protective geotextile does not the local strain to reach this level. Long-term durability is then a matter of correct materials selection and design.

**Summary**

• Coarse fills containing large or sharp stones can lead to holes and cuts during the installation of a geosynthetic.

• Correct installation procedures are essential to minimise this damage.

• For applications such as filters and separators robustness classes have been defined to enable selection of a geosynthetic that will resist damage.

• For reinforcements the loss of strength due to installation damage is taken into account by use of a reduction factor, which is measured by performing simulated site tests. This factor also covers the long-term performance.

• Dynamic loading (fatigue) can limit lifetime: its effect is measured by extrapolating the durations (or numbers of cycles) of short-term tests at higher loads to longer durations at lower loads.

• Resistance to abrasion is assured by suitable selection of materials based on short-term tests.

• Protection of geomembranes is assured by selecting a geotextile that limits the strain in the membrane.

# References

*Chapter 3.6*

• Brummermann K. Geomembranes under punctiform load. In Advanced Landfill Liner Systems, ed. August H et al, Thomas Telford Publishing, London, pp 251-260.

• Ehrler P, Gündisch W. Wirkung eingelagerter Teilchen auf das Kraft-/Dehnungsverhalten von Geotextilien. 6. Informations- und Vortragstagung über Kunststoffe in der Geotechnik, Munich, Germany, ed. Floss R; Geotechnik, 1999, special issue, pp 207-216.

• FGSV. Merkblatt über die Anwendung von Geokunststoffen im Erdbau des Straßenbaues, FGSVKöln, 2005.

• Greenwood, J H. The effect of installation damage on the long-term design strength of a reinforcing geosynthetic. Geosynthetics International, 2002, Vol. 9, No. 3, pp 247-258.

• Hufenus R, Rüegger R, Flum D, Sterba I J. Strength reduction factors due to installation damage of reinforcing geosynthetics. Geotextiles and Geomembranes, Vol. 23, 2005, pp 401-424.

• Koerner G R, Koerner R M. The installation survivability of geotextiles and geogrids. 4th International Conference on Geotextiles, Geomembranes and Related Products, The Hague, 1990. Balkema, Rotterdam, 1990, pp 597-602.

• Paula A M, Pinho-Lopes M, Lopes M L. Combined effect of damage during installation and long-term mechanical behavior of geosynthetics. EuroGeo4, Edinburgh, UK, paper 185 (2008).

• Müller W. HDPE geomembranes in geotechnics. Springer, Berlin, Germany, 2001.

• Retzlaff J, Müller-Rochholz J, Klapperich H, Böhning M. The behaviour of geosynthetics under cyclic load. EuroGeo4, Edinburgh, UK, paper 179 (2008).

• Zanzinger H, Hangen H, Alexiew D. Lifetime prediction of PET geogrids under dynamic loading. Kyushu 2007.

## 3.7    Durability of reinforcement and drainage applications using reduction factors

J.H. Greenwood

### 3.7.1    Reduction, safety and default factors

As indicated in chapter 2.3, reduction factors are a convenient method of incorporating durability aspects into certain types of design, notably for soil reinforcement. They provide a method by which the effects of different degrading factors on a key property can be combined into a single number. This can then either be used in design, or they can be compared with a limit of acceptability, leading to a pass-fail decision.

In soil reinforcement the key property is strength. The strength of the virgin geosynthetic is reduced by dividing by individual factors representing the effects of load, damage and environmental degradation, each of them related to the design life and temperature, to yield the maximum load that can safely be applied to the geosynthetic in practice. This is used in design to guide the type and quantity of material used in the structure. The same approach can be used other applications. For drainage the key property is flow in the plane, and the factors proposed are described in section 3.7.7. For a separator or a filter layer, the key property might be permeability or puncture resistance, and although attempts have been made to apply the same approach (see Case 4 and 5 in section 1.7.10) this is limited by general lack of data. Note also that the factors apply to only one key property. In the case of reinforcement they apply to strength but not to strain, for which a separate prediction must be made of the strain at the end of the design life (see section 3.7.6). It may also be more appropriate to predict oxidation in the form of a lifetime than as a reduction factor (see section 3.7.2.5). For seismic applications tests are necessary to predict the response to a dynamic load after a long period under static load (see section 3.7.5). In the case of drainage the key property is flow in the plane, but separate predictions are necessary for the time to collapse and the time to oxidation (see section 3.7.7.2). There are three types of factor:

**Reduction factors** represent the predicted change in a property for the anticipated design life and service conditions.**Safety factors** are used to cover an uncertainty. They may be calculated statistically or, if there is insufficient information for this, they will be estimated. **Default factors** based on generic information may be used where no information on the specific geosynthetic is available. Default factors are deliberately pessimistic.Convention states that the factors are set greater than unity and that the virgin property is divided by them. A factor of 1.0 is used where no change is predicted. Where a property is, for example, divided by 1.25 it would be equally logical to multiply it by the reciprocal, 0.8, but the convention has to be maintained to avoid confusion.

The use of separate safety factors implies that the effects of the different forms of degradation, for example load and chemical attack, are independent of one another. Where there is synergy between them, such as between temperature and load or temperature and chemical attack, the factor must be calculated to take into account both effects simultaneously.

Although the term 'safety factor' is sometimes used to include reduction factors, the difference between them is important. A single safety factor is used in exactly the same way as a reduction factor, but when two or more are present they should be combined differently to avoid an unnecessarily cautious result. This is explained further in section 3.7.3.

## 3.7.2 Procedure for determining the reduction factors for creep-rupture of reinforcements

### 3.7.2.1 Table of factors

The first objective for reinforcements is to provide an assurance that the reinforcement will not rupture during the design life of the structure. The second is to ensure that it will not extend or relax beyond unacceptable limits. As was explained in 2.3.6, the prediction of rupture is a good example of the use of reduction factors. The prediction of extension (creep strain) or stress relaxation is an example of a different approach and is discussed in section 3.7.6.

The selection of characteristic or initial strength of the geosynthetic is described in section 2.3.6. According to ISO/TR 20432 the characteristic strength of the geosynthetic is reduced by four reduction factors and one safety factor. These are shown in table 3.6 with their Dutch, German, US and British equivalents.

Table 3.6 Reduction factors according to ISO/TR 20432 and the Dutch, German, US and British equivalents.

| ISO TR 20432, USA | Netherlands, Germany | United Kingdom (BS8006) | Form of degradation |
|---|---|---|---|
| | | $fm_{111}$ | variability of initial strength |
| | | $fm_{112}$ | metallic reinforcement only |
| $RF_{CR}$ | $A_1$ | $fm_{113}$ | creep-rupture |
| $RF_{ID}$ | $A_2$ | $f_{m211}$ (short term) $f_{m212}$ (long term) | installation damage |
| | $A_3$ | | joints and connections |
| $RF_{WE}$ | $A_4$ | $f_{m22}$ | weathering |
| $RF_{CH}$ (USA: $RF_D$) | | | chemical degradation |
| | $A_5$ | | Special conditions, e.g. dynamic loads |
| $f_s$ | $\gamma_M$ | $f_{m122}$ (creep extrapolation only) | |

In ISO/TR 20432 there is no factor to cover the variability in the strength of the geosynthetic, since this is covered by the use of characteristic strength as the basic tensile strength. Where no value is available for the characteristic strength use the nominal strength quoted by the manufacturer, provided that he is covered by a suitable quality assurance system. The BS8006 allows an extra factor $fm_{111}$ for variation should the characteristic strength not be used.

These factors of safety cover those forms of degradation that are considered most likely for reinforcements. Solvents, which could swell the polymer and affect its creep strain, should not be present in any significant quantity outside waste containments. Environmental stress cracking is not regarded as likely in highly oriented polymers. The effects of freezing and thawing depend on the soil and its level of saturation and cannot be assessed for the geosynthetic in isolation. The Netherlands and German systems provide extra reduction factors for the strength of joints and connections ($A_3$) and for dynamic loading ($A_5$). These do not form part of the ISO system but are discussed in 3.7.2.8.

### 3.7.2.2   Calculation of $RF_{CR}$

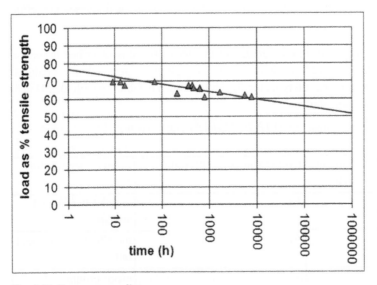

Fig. 3.81 Creep-rupture diagram.

Figure 3.81 shows a typical creep-rupture diagram for a specific temperature extended to 1 000 000 hours or 114 years. There should be at least 12 tests at a minimum of four different loads, measured in the principal direction of reinforcement using the same grips and specimen width as are used for measuring tensile strength. For conventional tests at least four values of the time to rupture should lie between 100 and 1000 h, four between 1000 and 10000 h and at least one should exceed 10000 h.

Fit a regression line as described in 3.4.3.3 and derive $RF_{CR}$ for the specified design life.

In figure 3.81, the load corresponding to a lifetime of 114 years is predicted to be 51.4% of tensile strength. The reduction factor is expressed as the inverse of the percentage above. For the above example, $RF_{CR}$ equals $1/51.4\% = 1.95$.

The rules for *incomplete tests* and the fundamental conditions for the validity of the extrapolation are explained in section 3.4.3.3 and the method for time-temperature shifting in section 3.4.3.5.

If some of the rupture points are from *temperature accelerated or SIM tests*, then ISO TR 20432 specifies a procedure and conditions which are set out in section 3.4.3.2. Provided that the conditions stated there are fulfilled, the SIM data may be combined with the conventional data and used to determine $RF_{CR}$. If not, $RF_{CR}$ should be determined from data from conventional testing alone, which will require additional tests.

An advantage of temperature accelerated or SIM tests is that they provide thermal shift factors as shown in figure 3.43. Where *the design temperature is lower than the test temperature*, as is often the case in Northern Europe, the graph can be extended to lower temperatures to derive a shift factor and applied to extend the predicted time to rupture (see section 3.4.6). For example, for a design temperature of 15 °C the shift factor in figure 3.44 would be about -0.9. When the regression line in figure 3.7.1 is shifted to the right by 0.9 decade (i.e. the durations are multiplied by $10^{0.9} = 7.9$) the design strength rises to about 55% and RFCR reduces to 1.82.

Should a *range of products* be subject to approval, a typical procedure would be to test a mid-range product provided that the materials used are identical and that the physical construction of the products is similar. In addition, further tests are performed on the lowest product and two on the highest product in the range. Provided that these lie within the 95% confidence limits of the results for the mid-range product then the creep data may be taken as representative for the product range.

In comparing values of $RF_{CR}$ do not forget the *uncertainty of the prediction*, described in 3.7.3. The minimum factor of safety is given there as 1.1 (i.e. 10%) and could be much greater. It is ridiculous to see manufacturers competing over differences of a few percent in $RF_{CR}$, differences that are almost certainly less than the uncertainty of the prediction.

In the absence of specific data an overall *default factor* of $RF_{CR} = 2.0$ for polyester and $RF_{CR} = 4.0$ for polypropylene reinforcements have been proposed (Greenwood and Shen 1994, Koerner 1997). The summary of international recommendations compiled by Zornberg and Leshchinsky (2001) gave the following default factors:

| Polyester | 1.5 to 2.5 |
|---|---|
| Polypropylene | 4.0 to 5.0 |
| Polyethylene | 2.5 to 5.0 |
| Aramid | 2.5 |
| Polyamide | 2.5 |

Further factors are listed in Bathurst et al. (2012).

### 3.7.2.3   Calculation of $RF_{ID}$

Determine $RF_{ID}$, the reduction factor for installation damage, should be deduced from site damage tests as described in chapter 3.6.4. $RF_{ID}$ is the strength of undamaged or control material divided by the strength of the damaged material.

If measurements are not available for the specific material and soil, $RF_{ID}$ should be estimated from existing data. Generally a table of values will be given according to the strength or mass per unit area of the product and the mean particle size, an example of which is shown in table 3.7.

Table 3.7 Reduction Factors $RF_{ID}$ for a typical Polyester Geogrid.

| Soil type | d90 particle size mm | Grade | $RF_{ID}$ |
|---|---|---|---|
| Sand A | < 4 | 35/20-20 5/30-20 | 1.17 1.06 |
| Sand B | < 2 | 80/20-20 110/30-20 | 1.03 1.04 |
| Sandy gravel A | < 12 | 35/20-20 55/30-20 | 1.17 1.09 |
| Sandy gravel B | < 8 | 80/20-20 110/30-20 | 1.13 1.07 |
| Coarse gravel A | < 65 | 35/20-20 55/30-20 | 1.29 1.19 |
| Coarse gravel B | < 40 | 80/20-20 110/30-20 | 1.19 1.18 |

Values of $RF_{ID}$ are naturally subject to scatter and it is not unusual for there to be a few unexpectedly large or small values in the table. In such cases the values should be adjusted – for example, by fitting an empirical function - to give a smooth transition from fine to coarse soils and from lighter to heavier products.

If the life prediction is to be performed on a geosynthetic from a range of products for which $RF_{ID}$ has been performed on both heavier and lighter grades within the range, then $RF_{ID}$ can be deduced by interpolation (figure 3.82). This is provided that there is a clear relation between $RF_{ID}$ and weight, tensile strength or, for coated geogrids, coating thickness, for the products tested. If $RF_{ID}$ is only available for a lighter grade, then this value should be used. The use of a value of $RF_{ID}$ measured on a heavier grade is not permitted.

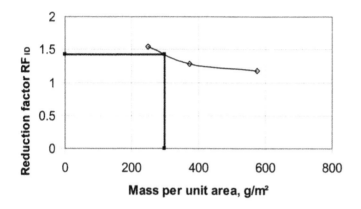

Fig. 3.82 $RF_{ID}$ plotted as a function of mass per unit area. For a 300 g/m² material $RF_{ID}$ is interpolated to be 1.4.

If tests are not available for the soil for which a life prediction is to be made, but where $RF_{ID}$ is known for the same geosynthetic in two other soils, one with a larger and one with a smaller grain size, $RF_{ID}$ can also be calculated by interpolation (figure 3.83) Be extra cautious with soils containing a wide particle size distribution or very angular particles. If $RF_{ID}$ is known only for a soil with a larger grain size which is likely to be more aggressive, then this value should be used for the new soil. The use of $RF_{ID}$ measured using a soil with a smaller grain size is not permitted.

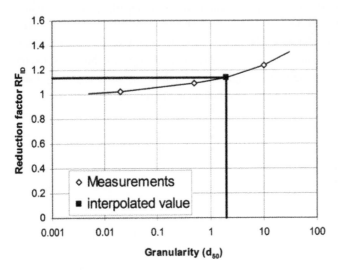

Fig. 3.83 $RF_{ID}$ plotted as a function of soil granularity. Note the logarithmic scale. For $d_{50} = 2$ mm $RF_{ID}$ is interpolated to be 1.13.

For example in table 3.7 the product 80/30-20 has been measured as having $RF_{ID} = 1.03$, 1.13 and 1.19 for soils with maximum $d_{90}$ of 2, 8 and 40 mm respectively. For the other three soils proceed as follows. For the soil with maximum $d_{90} = 4$ the interpolated value of $RF_{ID}$ would be 1.09 and for the soil with maximum $d_{90} = 12$ the interpolated value of $RF_{ID}$ would be 1.13. The coarsest soil in table 3.7, maximum $d_{90} = 65$, is beyond the range of interpolation and a further test would be required to define $RF_{ID}$. For the product 35/20-20, where no value is given for the lightest sand with maximum $RF_{ID} = 2$, use the value for the next lightest sand, i.e. $RF_{ID} = 1.17$. It is not permissible to estimate a lower value for $RF_{ID}$.

Derivation of $RF_{ID}$ from measurements made using ISO 10722 was validated by Hufenus et al (2005) for a range of materials but is not generally accepted.

Some tables have been issued of general values of $RF_{ID}$ based on a comprehensive series of tests (Ref: Elias 1997, Hufenus et al. 2002). These are shown tables 3.8 and 3.9 More comprehensive tables are given in Hufenus et al. (2005). The user is advised to refer to the original documents to confirm that the materials and soils used were sufficiently comparable for the materials for which a life prediction is to be made.

Table 3.8 Reduction factors RFID from Elias (1997).

| Geosynthetic | No 1 Backfill Max. Size 102 mm d50 about 30 mm | No 2 Backfill Max. Size 20 mm d50 about 0.7 mm |
|---|---|---|
| HDPE unixial geogrid | 1.20 – 1.45 | 1.10 – 1.20 |
| PP biaxial geogrid | 1.20 – 1.45 | 1.10 – 1.20 |
| PVC coated PET geogrid | 1.30 – 1.85 | 1.10 – 1.30 |
| Acrylic coated PET geogrid | 1.20 – 2.05 | 1.20 – 1.40 |
| Woven geotextiles(PP & PET)* | 1.40 – 2.20 | 1.10 – 1.40 |
| Nonwoven geotextiles (PP & PET* | 1.40 – 2.50 | 1.10 – 1.40 |
| Slit film woven PP geotextile | 1.60 – 3.00 | 1.10 – 2.00 |

* Minimum weight 270 g/m$^2$

Table 3.9 Reduction factors RFID from Hufenus et al. (2002).

| Geosynthetic | Fine-grained soil | Rounded coarse-grained soil | Angular coarse-grained soil |
|---|---|---|---|
| Uniaxial HDPE grids | 1.0 – 1.1 | 1.1 – 1.2 | 1.2 – 1.4 |
| Biaxial PP grids | 1.0 – 1.2 | 1.1 – 1.3 | 1.2 – 1.5 |
| PET flat rib grids | 1.0 – 1.1 | 1.0 – 1.1 | 1.0 – 1.1 |
| Coated PET grids | 1.0 – 1.1 | 1.1 – 1.2 | 1.2 – 1.3 |
| PP & PET wovens | 1.0 – 1.2 | 1.1 – 1.4 | 1.2 – 1.5 |
| PP & PET nonwovens | 1.0 – 1.1 | 1.2 – 1.4 | 1.3 – 1.5 |
| PP slit tape wovens | 1.0 – 1.2 | 1.1 – 1.3 | 1.2 – 1.4 |

Recently Bathurst et al. (2011) have performed a detailed reliability analysis based on a wide range of reported values of $RF_{ID}$. The values are listed in table 3.10: for other details including variability refer to the original paper. Note that since these are test results some strengths for exposed materials turn out less than those of unexposed materials: there are physical explanations for this, but the values of RFID used in design should not be less than 1.0.

Table 3.10 Reduction factors $RF_{ID}$ from Bathurst et al (2011a).

| | $RF_{ID}$ for d50 < 19 mm | $RF_{ID}$ for d50 > 19 mm |
|---|---|---|
| HDPE uniaxial geogrids | 0.99 to 1.17 | 1.09 to 1.43 |
| PP biaxial geogrids | 0.94 to 1.11 | 0.97 to 1.45 |
| PVC coated PET geogrids | 0.95 to 1.39 | 1.07 to 1.85 |
| Acrylic and PP coated PET geogrids | 1.05 to 1.37 | 1.48 to 2.02 |
| Woven geotextiles | 0.89 to 1.66 | 1.20 to 4.93 |
| Nonwoven geotextiles | 1.06 to 1.46 | 1.74 to 4.96 |

*Note the large variation in these factors* even for the same type of product. When a design is made based on one brand of product normally the factors for that product have been used. *If at the tender stage the contractor proposes to use a different (normally cheaper) product these factors need to be checked thoroughly before that product is approved.* One might even go so far that the structure needs to be redesigned using the proper factors for the alternative product.

Only the damage caused during installation is referred to here. The possible long-term effects of such damage are discussed in 3.6.6. In reinforcement applications damage due to dynamic loading is often regarded as negligible, but if it is likely to be significant, for example under railway ballast, then it should be handled as described in 3.6.7.

Installation damage is treated as being independent of temperature.

### 3.7.2.4    Calculation of $RF_W$

Weathering is discussed in chapter 3.3. Since most geotextiles are covered on site, no long-term prediction is required for weathering. For reinforcement applications a reduction factor is required only if there is a possibility of degradation during installation, in the same manner as for installation damage.

The only estimates of $RF_W$, the reduction factor for weathering, are those in Annex B common to EN 13249-13257 and 13265, table 3.11.

Table 3.11 Definition of $RF_W$ according to Annex B common to EN 13249-13257 and 13265.

| Retained strength after testing to EN 12224 | Time allowed for exposure on site | Reduction factor $RF_{WE}$ |
|---|---|---|
| >80% | 1 month | 1/percentage retained strength |
| 60% to 80% | 2 weeks | 1.25 |
| <60% | 1 day | 1.0 |
| Untested material | 1 day | 1.0 |

For example, if the retained strength after testing to EN 12224 was 87%, then $RF_{WE} = 1.15$. If the retained strength is greater than 95%, then $RF_{WE}$ may be set as equal to 1.0. No such procedure has been specified for tests to ASTM D4355.

As before, if appropriate information is not available for one product within a range then it is acceptable to use the results for a lighter grade, since this is more likely to be sensitive to weathering. Values from heavier grades should not be used.

The condition that the material is to be covered in service is critically important. In fact, most reinforcements are covered. Sometimes, however, it is unavoidable that the geosynthetic is left exposed for longer than one month, in which case longer tests using the same methods as in EN 12224 and ASTM D4355 should be performed to ascertain whether there is any significant reduction in strength. Geotextiles are not generally exposed to light permanently as is the case for geomembranes at the edges of reservoirs, canals and containment ponds, for which it has so far not been possible to establish short-term test methods or criteria to assure long-term resistance to light.

### 3.7.2.5   Calculation of $RF_{CH}$

Predicting the reduction in strength arising from chemical and environmental effects is without doubt the most difficult aspect of life prediction for reinforcements. Regular extractions from site over a number of years may lead to a graph of strength plotted against time that can be extrapolated to the design life. This is the ideal, since no laboratory test can simulate precisely the environment surrounding the geosynthetic. It assumes, however, that the change in strength is gradual and predictable and not sudden, as can occur in polyolefins. It also assumes that the environment stays unchanged throughout the life of the product. However, information of this type is not available, and will only be in the future if proper preparations are made now for extracting samples on site (see section 3.8.5). Therefore the loss in strength due to environmental degradation must be predicted from accelerated testing. To do this we require:
- a graph of the loss of strength with time for each temperature;
- for each temperature the rate of loss of strength (or other property) or the time to a set end point such as 50% loss of strength;
- a formula or graph – generally an Arrhenius diagram – to enable us to move from one temperature to another.

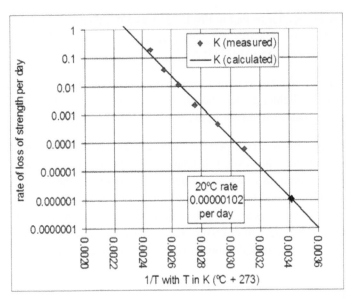

Fig. 3.84 Arrhenius diagram showing the rate of loss of strength plotted against the inverse of temperature in K.

Figure 3.84 shows a typical Arrhenius diagram of rate of loss of strength plotted against the inverse of the temperature in K, with a regression line fitted. Extend this line until it reaches the inverse of the design temperature, here $0.0034 = 1/(273 + 20)$ K (a similar deduction could be made for a lower design temperature). Read off the corresponding rate of loss of strength, which is 0.00000102 per day (it is less confusing to present such low rates in absolute terms rather than as a percentage). Multiply this by the design life (in the same units of time) to give the predicted loss in strength. For example, in this case the loss of strength after 100 years or 36524 days will be 3.8% and the retained strength will be 96.2%.

$RF_{CH}$ is the inverse of the percentage retained strength.

In this example $RF_{CH} = 100/96.2 = 1.04$.

The use of rate of loss of strength assumes that this is constant, i.e. that a graph of loss of strength against time would be a straight line. If the Arrhenius diagram refers to a duration rather than a rate, such as time to 50% loss of strength, then the extrapolated diagram (which will slope upwards) will give the time to 50% loss of strength at the design temperature. If the rate of loss of strength is not constant, the following procedure is proposed. Fig 3.85 (a) shows the reduction in strength over a series of times at one temperature, as will have been performed to establish one point on the Arrhenius diagram, in this case 50% after 190 h. The Arrhenius diagram is assumed to predict a lifetime of 250 years at the service temperature. Figure 3.85 (b) shows the same curve as in 3.85 (a) scaled to pass through the point (250, 50%). Using the same scale, read off the residual strength predicted after the design life of 100 years, which is 90.3%. $RF_{CH}$ should then be set equal to $1/90.3\% = 1.11$.

276

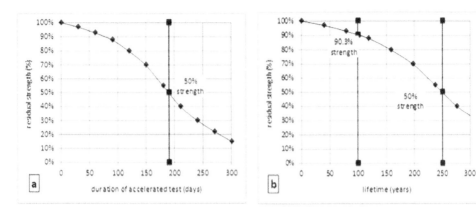

Fig. 3.85 (a) Curve of loss of strength against time, (b) scaled such that it passes though the point (250 years, 50%). The predicted retained strength after 100 years is then 90.3%.

If the appropriate information is not available for one grade of a product range then apply the same rule as for installation damage and weathering: use the value of $RF_{CH}$ for a lighter grade. Data for a heavier grade should only be used if it can be argued that the rate of degradation is the same, for example that the filaments are of similar size and that the thickness of the product does not inhibit access by the chemical environment – including oxygen – to the surface.

In some cases, more particularly for oxidation, degradation takes place over a relatively short time following a long incubation period (see Mode 3 in chapter 2.3 and chapter 3.1). In this case $RF_{CH}$ will remain close to 1.0 during the incubation period and then fall rapidly to zero. It may be better to make a separate calculation of the time to failure instead of applying a reduction factor (see section 1.6.4). This time to failure, modified by an appropriate safety factor, can then be compared with the design life.

### 3.7.2.6   $RF_{CH}$ for polyesters

Chapter 3.2 describes the methods for predicting the loss of strength of polyesters due to hydrolysis. ISO TR 20432 introduces simplified procedures for polyester geosynthetics which satisfy certain basic criteria, assuming no post-consumer or post-industrial recycled material is used.

For polyester either:
- The polyester geosynthetics used for reinforcement, or the yarns from which they are made, should exhibit no more than a 50 % reduction in strength when subjected to EN 12447.
- Or: The CEG measured according to GRI-GG7 should be less than 30 meq/g, and the number averaged molecular weight, Mn, determined according to GRI-GG8, should be 25,000 or more. Both criteria should be satisfied.

For a geosynthetic that satisfies either recommendation used in saturated soil, $RF_{CH}$ is given in table 3.12. These numbers are based on several independent studies including accelerated testing. The interpretation tends towards the conservative. They have, however, not yet been validated by observations on material taken from sites where the environment has been sufficiently monitored.

Table 3.12 Default values of $RF_{CH}$ for polyester reinforcements from ISO TR 20432 at 25 °C.

| pH range | Design life (years) | Service temperature (°C) | $RF_{CH}$ |
|----------|---------------------|--------------------------|-----------|
| 4 - 9 | 25 | 25 | 1.0 |
| 4 - 8 | 100 | 25 | 1.2 |
| 8 - 9 | 100 | 25 | 1.3 |
| 4 - 9 | 25 | 35 | 1.4 |

These values are given for the maximum European soil temperature of 25 °C. They are necessarily conservative for most soils in Europe: in the UK the average soil temperature at depth is approximately 10 °C. Assuming an activation energy of 105 kJ/mol, 25 years at 25 °C corresponds to 109 years at 15 °C and 236 years at 10 °C. Note, however, that due to the exponential relation warmer days will accelerate degradation more than cooler days retard it, so that an 'upper average' soil temperature should be chosen in preference to a true mean.

If the soil is totally dry, such as occurs (most of the time) in desert regions then there is no possibility of hydrolysis and $RF_{CH}$ = 1.0, excluding any other possible types of degradation. For partially saturated soils one can make a linear interpolation according to the relative humidity. For example, for a soil with pH 4 – 8 at 25°C the table above gives $RF_{CH}$ = 1.2 for 100 years for a saturated soil while $RF_{CH}$ = 1.0 for a dry soil. For a soil with 70% relative humidity $RF_{CH}$ can be derived as 70% × 1.2 + 30% ×1.0 = 1.14. In case of any doubt about the environment, however, the figure for a saturated soil should be used. Some caution should be applied to these factors which have not been validated.

In some cases more favourable predictions are proposed as a result of Arrhenius tests on a particular grade of polyester fibres. The assessor will have to decide whether these predictions can be maintained for this product or whether it is more circumspect to apply the default factors which are based on a wider range of independent tests. As mentioned in section 3.7.2.2 in the case of RFCR, it is ridiculous to see manufacturers competing over differences of a few percent in RFCH, differences that are almost certainly less than the uncertainty of the prediction.

### 3.7.2.7   $RF_{CH}$ for polypropylene and polyethylene

Chapter 3.1 describes the methods for predicting the change in strength with time for polyethylene and polypropylene geosynthetics, including:
- Simple oven ageing tests using the method of ISO 13438 Methods A2 or B2 extended over a range of temperatures and times with a defined retained strength as the criterion.

In this case a chart of the rate of reduction in strength against time will be required to predict the retained strength under for the design life and service temperature as in figures 3.88 and 3.89.

- Oven ageing coupled with OIT tests to define two or more stages of oxidation, in which the lifetime will consist of one or more incubation stages followed by the stage in which the strength diminishes. In this case the design life must take into account the durations of the incubation stages and the retained strength calculated for the time remaining. If the incubation phases have not been completed by the end of the design life then there is no predicted reduction in strength.
- The pressurized oxygen test using ISO 13438 Method C1 or C2 but with a range of temperatures and pressures. The reduction in strength following exposure to these conditions will be charted on a three dimensional diagram which will then be extrapolated to the service conditions to yield a predicted retained strength.

All these methods are designed to predict a retained strength for a defined design life and environment. Then:

$RF_{CH}$ = initial strength/predicted retained strength.

ISO TR 20432 sets out default factors for polypropylene and polyethylene, namely that if the material passes ISO 13438 Method A2 for polypropylene or B2 for polyethylene, or C2 for either, then $RF_{CH}$ = 1.3 for a design life of 100 years at a temperature of up to 25 °C. This default factor is not based on the same quantity of evidence as the factor for hydrolysis, particularly when one considers the wide variety of products made from these polymers, and should be treated with great caution.

### 3.7.2.8 German factors $A_3$ and $A_5$

The German factor $A_3$ takes into account the reduction in strength due to joints and connections, either between sections of geosynthetics or, more commonly, where the end of the geosynthetic is fixed to a wall, block or anchor. The simplest procedure is to assume that the reduction in long-term strength is equal to the reduction in short-term strength as measured in an appropriate tensile test. The reduction factor is then the tensile strength of the virgin geosynthetic divided by the tensile strength of the joint. This procedure assumes that the joint does not have a disproportionately low long-term strength due, for example in an unpredicted change in the mechanism of rupture as occurs in plastic pipes under pressure. The only means of assurance is to perform creep-rupture tests on the joints themselves and to set up an appropriate creep-rupture diagram. $A_3$ is then the ratio of the creep-rupture strength for the virgin geosynthetic for the set design life divided by that for the joint.

Factor $A_5$ takes into account dynamic loading. The method of measurement and extrapolation is described in 3.6.7, resulting in a plot of dynamic load against number of cycles to failure, not of residual strength against time or number of cycles. To study the behaviour in detail, one could follow the methods described in section 3.4.5 for describing the response to seismic loads. If this amount of effort is not justified, a conservative approach might be to set

$A_s = 1.0$ but to regard the final design strength as the sum of the maximum permissible static and dynamic loads.

### 3.7.3   Uncertainties and factors of safety

#### 3.7.3.1   Introduction

Any prediction of lifetime will have an associated uncertainty. The most common way of allowing for this is by applying a factor of safety, although more sophisticated methods of reliability based design have recently been published (Bathurst et al. 2011, 2012). It is important to realize that any safety factor based on geosynthetic properties alone can only reflect the variability of the material. The uncertainties associated with the soil structure form part of the design have to be assigned separately by the designer; they may be considerably larger than those derived from the material properties alone (see section 2.3.7). For reinforcing geosynthetics the variation in tensile strength is accounted for by assuming a characteristic or minimum strength for the material, based on the variation in individual results. This should also reflect the variability of the test method itself. Note that in creep measurements the chief contribution to the level of variability is the stability of the test temperature.

From their analysis of a wide range of creep tests Bathurst et al confirm that the variability in extrapolated design strength based on creep-rupture and frequently on SIM tests is similar to the variability of the original strength of the product, i.e. that the use of characteristic strength in itself provides sufficient safety. The same is true of installation damage at lighter loads, and they conclude that it is more important to select a geosynthetic appropriate to the granularity of the soil and load applied, rather than design for a large value of RFID and the possibility of a high degree of variability.

It is important to apply only one factor of safety. A safety factor based on a 'two-sided' 95% confidence limit represents a 1 in 20 probability that the property will lie outside the confidence limits. The 5% "rogue" strengths are divided into 2.5% that are lower than the lower confidence limit and 2.5% that are higher than the upper confidence limit, which do not worry us. The factor therefore represents a 'one-sided' 97.5% probability (or 1 in 40) that the strength will be sufficient. If this is multiplied by a second safety factor also based on a 95% confidence limit, the probability of a low strength jumps from 1 in 40 to 1 in 1600 and the safety factor becomes ridiculously cautious. To retain the 95% level of confidence they should be combined as follows:

$$f_s = 1 + \sqrt{[(FS_1 - 1)2 + (FS_2 - 1)^2]}$$

Where one factor is much larger than the other, for example $FS_1 \gg FS_2$, it is sufficient to use the larger factor (here $FS_1$) alone.

### 3.7.3.2   Factors of safety in ISO/TR 20432

The factor of safety $f_s$ included in the calculation of design strength in ISO/TR 20432 accounts solely for the following:

- uncertainty of extrapolation of creep-rupture data ($R_1$)
- uncertainty of extrapolation of Arrhenius data ($R_2$).

These are explained in more detail in the following sections.

### 3.7.3.3   Definition of $R_1$

ISO TR 20432 specifies $R_1$ in a manner that depends on the degree of extrapolation. This is the ratio of the design life $t_d$ to the duration of the longest measured time to rupture after time-temperature shifting, denoted as $t_{max}$. It is therefore an advantage to extend testing to have $t_{max}$ as large as possible. tmax may refer to an incomplete test provided that the test satisfies the condition in section 3.4.3.3 and is included in the calculation of creep-rupture. The rules are:

If $t_d / t_{max}$ <10, $R_1$ = 1.0.

If $10 < t_d / t_{max}$ <100, $R_1$ = $1.2^{r-1}$ where $r = \log_{10} (t_d / t_{max})$.

$R_1$ increases with $r$ from 1.0 for extrapolation by a factor of $t_d / t_{max}$ = 10 to 1.2 for extrapolation by a factor of $t_d / t_{max}$ = 100. No rules are set for when $t_d / t_{max}$ >100. If $t_d$ = 50 years and $t_{max}$ = 7791 h = 0.89 years as in figure 3.7.5, then $r = \log_{10} (50/0.89)$ = 1.75 and $R_1$ = $1.2^{r-1}$ = 1.15.

BS 8006:1996, now superseded, set out a larger factor of safety to allow for extrapolation, rising from 1.0 for when the design life is 10 times the test duration to 2.0 when the ratio is 100 times. This extra caution was introduced to allow for a possible knee in the creep-rupture of extruded polyethylene geogrids. Improvements in materials indicate that this is now unlikely to occur in commercially available geogrids. The effect of $R_1$ on the design strength as calculated according to ISO TR 20432 and BS8006:1996 is shown in figure 3.90.

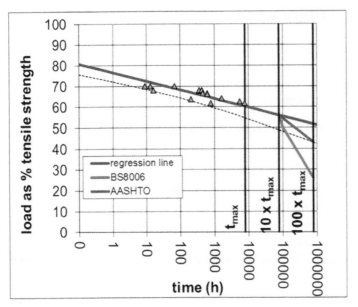

Fig. 3.86 Effect of the contributory factor R$_1$ in reducing the design strength when extrapolation exceeds a factor of 10, as defined by ISO TR 20432 and BS8006:1996 (superseded).

### 3.7.3.4   Definition of R$_2$

If default factors are used for RF$_{CH}$ (see section 3.7.2.6) then ISO TR 20432 defines $R_2$ = 1.0. If RF$_{CH}$ is based on an Arrhenius diagram, then $R_2$ is calculated from the upper 95% confidence limit of that diagram. The upper confidence limit corresponds to a faster rate of degradation.

Figure 3.87 shows an Arrhenius diagram plotted as the rate of loss of strength (logarithmically) as y against the inverse of absolute temperature $1/\theta$ as $x$. The equation of the fitted straight line is

$$y = y_{mean} + S_{xy}/S_{xx}(x - x_{mean})$$

where

$x_{mean}, y_{mean}$ are the mean values of $x$ and $y$ respectively

$S_{xx} = \Sigma(x - x_{mean})^2$

$S_{yy} = \Sigma(y - y_{mean})^2$

$S_{xy} = \Sigma(x - x_{mean})(y - y_{mean})$

The equation of the upper confidence limit is given in section 2.3.9.9.

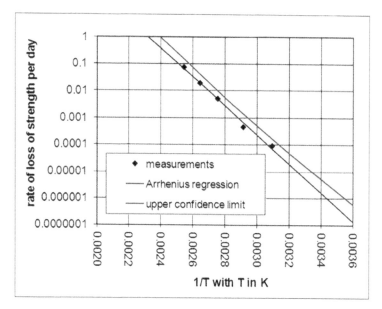

Fig. 3.87 Arrhenius diagram showing upper confidence limit.

The rates of loss of strength for the design life are read both from the Arrhenius curve and the upper confidence limit. The respective retained strengths are then calculated. $R_2$ is defined most simply as their ratio, that is the rate of the retained strength predicted by the upper confidence limit divided by the lower value predicted by the Arrhenius curve. Thus $R_2 > 1$.
In figure 3.87 the rate of loss of strength at 21°C ($1/T = 0.0034$) is 0.00000167 per day, giving a retained strength of 94% after 100 years and $RF_{CH} = 1.06$. The rate of loss of strength according to the upper confidence limit is 0.00000537 per day, giving a retained strength of 80% after 100 years. $R_2 = 94/80 = 1.175$.

### 3.7.3.5  Calculation of fs

$R_1$ and $R_2$ are then combined together to give the factor of safety fs. This is performed according to the following statistical formula, which maintains the 95% level of confidence:

$$f_s = 1 + \sqrt{\{(R_1 - 1)^2 + (R_2 - 1)^2\}}.$$

| |
|---|
| For example:<br>$R_1 = 1.15$<br>$R_2 = 1.175$<br>$f_s = 1 + \sqrt{\{(0.15)^2 + (0.175)^2\}} = 1.23$ |

If $R_1$ or $R_2$ is significantly less than the other then it can be ignored and $f_s$ can be set equal to the larger factor.
$R_1$ depends on the duration of the longest creep test, $t_{max}$. If the creep data are extrapolated by less than $10 \times t_{max}$ then $R_1 = 1.0$. For extrapolation to design lives td between $10 \times t_{max}$ and

283

$100 \times t_{max} R_1$ increases with log ( $t_d / t_{max}$ ) from 1.0 to 1.2. Further details are given in section 3.7.3.3.

The default value for $f_s = 1.2$.

Typical values of $f_s$ in the summary of international recommendations compiled by Zornberg and Leshchinsky in 2001 lie between 1.1 and 1.7, but it is not clear what uncertainties these values are intended to cover.

A minimum value is recommended of $f_s = 1.2$.

### *3.7.4    Calculation of the overall reduction factor RF*

In summary, the above sections have shown how to define:

- $RF_{CR}$ : creep-rupture
- $RF_{ID}$ : installation damage
- $RF_{WE}$ : weathering
- $RF_{CH}$ : chemical degradation

The general reduction factor RF is calculated as

$$RF = RF_{CR} \times RF_{ID} \times RF_{WE} \times RF_{CH} \times f_s.$$

The characteristic or nominal strength is divided by RF to give the long term design strength.

---

For a characteristic strength of 80 kN/m, $RF_{CR} = 1.28$, $RF_{ID} = 1.64$, $RF_{WE} = 1.0$, $RF_{CH} = 1.2$ and $f_s = 1.23$, RF = $1.28 \times 1.64 \times 1.0 \times 1.2 \times 1.23 = 3.10$ and the design strength = 80/3.10 = 25.8 kN/m.

---

RF is divided into the characteristic strength of the reinforcement, to give the design strength in kN/m as shown. This strength, represents the highest load which, if applied on the day of construction, is predicted not to lead to rupture within the design life. Applied to the characteristic strength, it represents a lower confidence limit in the same manner as the characteristic strength itself.

Creep strain should be presented separately as isochronous curves as described in 3.7.6.

### *3.7.5    Residual strength and the response to seismic events*

Following the success of reinforced slopes in surviving large earthquakes such as the Great Hanshin-Awaji or Kobe earthquake of 1995, much work has been devoted to seismic design of geosynthetic reinforced structures. From a life prediction aspect the principal interest is in any change in strength over time which could compromise the ability of the structure to resist a seismic event in the future. The reduction in strength due to damage, weathering and environmental effects is thus no different to that for normal reinforcement applications. The difference lies in the effect of sustained load on the future strength.

Section 3.4.5 explained how sustained load slowly reduces the strength until the geosynthetic is no longer able to support the applied load itself. Accelerated tests have indicated that this reduction in strength is very slow over most of the lifetime of the geosynthetic, falling rapidly at the end. The physical reason for this is not yet clear. If the applied load is expected to be lower than $RF_{CR} \times T_{char}$, it can be more appropriate to calculate the time to failure corresponding to the applied load and to check that this substantially exceeds $t_D$. On the basis of current measurements it may then be assumed that the strength remains close to Tchar over the design life. This is particularly relevant to seismic design and to other cases where a certain reserve strength has to be assured. The condition is, however, that the strength should be measured at an appropriate strain rate, which for seismic events is likely to be much faster than in a standard tensile test. Work by Hirakawa et al. has confirmed that the strength in response to a seismic event depends only on the immediate strain rate and not to the time the material has been under sustained load (see section 3.4.5).

Provided that the lifetime under load is a small fraction of the predicted time to rupture under that load, such as < 10%, then $RF_{CR}$ can be omitted and

$$RF = RF_{ID} \cdot RF_{WE} \cdot RF_{CH} \cdot f_s.$$

The modulus of the geosynthetic following a period under sustained load may have increased and will be dependent on strain rate as described in section 3.4.5.

### 3.7.6 Presentation of creep strain and stress relaxation data for design

It is not our purpose to comment on the methods used to calculate and design reinforced walls; only how to measure and predict the data that those methods require. Isochronous curves generally suffice to indicate the relation between strain, applied load and time. *Creep strain should therefore be presented in this manner.*

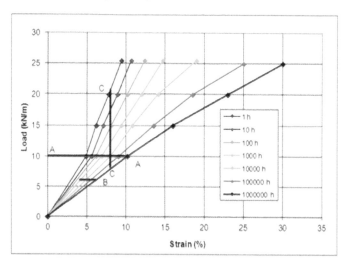

Fig. 3.88 Isochronous diagram of load plotted against total strain measured after specified durations.

Tests at at least four loads (typically 10%, 20%, 30% and 40% of tensile strength) are necessary to set up an isochronous diagram as in figure 3.88. If the design temperature is not 20 °C, adjust the durations on the isochronous lines as described in section 3.4.6.

The following options describe how to derive a maximum design load given a limit on creep or relaxation (see also section 3.4.2.10):

- Where further extension is possible the reinforcement will creep: in this case it may be specified that the total strain should not exceed a set limit. Where this *maximum total strain* is specified, read off the maximum load from the 1000000 h isochronous line. If in figure 3.88 the maximum total strain is 10% then line A-A shows that the design load for a 1000000 h design life would be 10 kN/m.
- If instead of the total strain the *maximum strain after construction* is specified, and the duration of construction and loading is 1000 h, determine the maximum load at which the difference between the 1000 h and 1000000 h isochronous lines is equal to this strain. For example if the maximum strain after construction is 2% then line B in figure 3.88 gives a design load of 6 kN/m. It may help to draw a separate diagram as in figure 3.89.
- Where further extension is prevented by anchoring the ends of the geosynthetic, the stress in the geosynthetic will relax. If it is specified that the *prestress in the geosynthetic should not fall below a certain minimum*, draw a vertical line to establish the stress on loading. In figure 3.88 line C-C shows that an initial load of 20 kN/m will fall to 8 kN/m after 1000000 h. Note however that after creep or relaxation the response to rapid loading will be different, as described in 3.4.5.
- If the geosynthetic can extend but is restrained by the soil there will be a *combination of limited creep and limited stress relaxation*. No precise rule can be given for this. The combined effect stress and the combined extension and relaxation (see section 3.4.1) will have to be estimated from the isochronous curves. The soil should not be allowed to exceed its own failure strain.

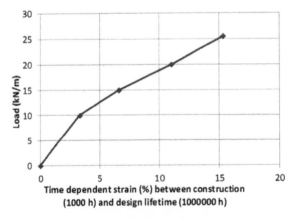

Fig. 3.89 Isochronous curve showing the strain developed between an assumed loading time of 100 h and a design life of 1000000 h (114 years).

The actual rate of loading during construction is likely to be very slow. The strain generated can be estimated from the isochronous curve corresponding to the loading period, as described above, or by performing tensile tests over a wide range of strain rates, establishing a relation between modulus and strain rate, and extrapolating this to the strain rate experienced during construction. Bathurst et al have demonstrated that measurements based on creep tests give the higher but converge at longer times.

Provided that the correct temperature is chosen, creep strain in geotextiles can be taken as insensitive to installation damage (except in severly damaged materials which would not be acceptable for use), weathering and most environmental effects. The creep of polyamides will be sensitive to humidity and that of polyesters may be to a limited extent, but there is insufficient data to define this. Design data can therefore be derived from an isochronous diagram adjusted, if necessary, to the design temperature, with no further reduction factors.

Note that the actual strains occurring in a construction have always been measured as much smaller than those predicted in the calculations. The main reason for this that in the calculations for the serviceability (or service limit state) the assumptions (in situ temperature, loads, construction strain, weights, etc) are still very pessimistic resulting in estimates of strain that are too high.

## 3.7.7  Drainage geosynthetics

### 3.7.7.1  General remarks

Drainage geosynthetics were described in chapter 2.2 and in the context of compressive creep in chapter 3.5. While their principal function is to allow water flow in the plane, they act simultaneously as filters. They must have sufficient mechanical strength to resist pressure from the soil or waste, they must be resistant to leachates and in some cases offer a level of mechanical protection.
Life prediction of drainage geosynthetics consists of two parts (Müller et al 2008):

* avoidance of flow stoppage due to collapse or oxidative degradation;
* maintenance of sufficient flow (transmissivity) over the design life after allowing for the effects of compressive creep, intrusion and clogging.

### 3.7.7.2  Collapse and oxidative degradation

Prediction of time to collapse of the core should be based on tests comprising the following:
* Perform compression tests over a range of compressive loads, augmented by shear if appropriate, and measure the time to collapse.
* Establish a diagram similar to figure 3.81.
* Extrapolate the fitted line from high loads and short times to lower loads and long times to predict the time to collapse under service conditions or, conversely, the maximum compressive load corresponding to the design life.

Further information is given in chapter 3.5. The tests should be performed at the service temperature, or tests should be performed over a range of temperatures to establish a shift factor as for tensile creep. Müller et al. developed a testing device which allowed testing of samples under a slope, in a test chamber at various temperatures.

Different cores behave very differently. Müller et al showed that a core consisting of a wide, flexible, wave-formed random array of extruded PP monofilaments deformed steadily under high pressure with no predicted instability, while one consisting of a rigid zig-zag array became unstable at a compressive deformation, which was further reduced by shear stresses and high temperatures, such as are present on the side slope of a landfill. Any such tests are specific to the type of drainage material tested.

Many drainage cores consist of polypropylene and polyethylene which degrade by oxidation leading to loss of strength. The material in the core may well have a low degree of orientation, be less comprehensively stabilized or contain recycled polymer. If the core degrades far enough it will collapse and obstruct flow. The methods for accelerated chemical degradation described in 3.1 are not necessarily suitable for determining RFCD for change in flow rate after exposure, not least because of the size of sample required, and it may be necessary to base predictions on a property such as tensile strength. The procedure is therefore:
- Predict the service life of a polypropylene or polyethylene core (and of the filter fabric if required) for oxidation as described in 3.8.1.

Müller et al. (2009) performed an extensive analysis of the oxidative behaviour of three different types of drainage materials, each comprising nonwovens and cores. They recommend oven ageing tests coupled with determination of the depletion of antioxidants, measured using OIT or chemical analysis, which is essential in the case of HAS stabilizers. The rate of depletion is described by an exponential formula and in some cases by two exponential terms superimposed. End of life is approximately 10% residual antioxidant, beyond which the strength begins to degrade, although the critical content is even less for phenolic and phosphitic antioxidants. The durations to this end point depend on temperature and can be fitted to an Arrhenius equation and extrapolated to service temperatures. This is regarded as more reliable than the extrapolation of mechanical strength where the activation energy can change at about 80 °C. Apart from the obvious factors such as temperature and fibre thickness, the rate of depletion was found to be unexpectedly lower for thick nonwovens, while water immersion led to the leaching and also hydrolysis of certain antioxidants. Note that a landfill environment may be hot and contain transition metal ions, which will accelerate degradation, but may also be depleted of oxygen, which will retard it.

### 3.7.7.3   Restriction of flow

The methods for predicting the reduction in flow due to compression creep are described in chapter 3.5. The FGSV Merkblatt (2005) proposes tests under normal loads of 20, 50, 100 and 200 KPa, with added shear when the application is a side slope.

Measurement of creep is however not sufficient on its own and must be complemented by tests to establish the relation between compressive strain and flow (Müller et al. 2008, Giroud et al 2000b). The effect of a soft bedding, which intrudes into the drainage space, must be considered either by performing tests using this bedding, or by applying a factor equal to the ratio of the transmissivities using soft and hard bedding measured in short-term tests. If temperature accelerated tests are used, a shift factor will be generated with which it is possible to make predictions for creep at lower soil temperatures.

Giroud et al. (2000a, 2000b) attempted to predict long-term water flow by a reduction factor approach as follows:

| $RF_{IMCO}$ | Reduction factor for instantaneous compression |
|---|---|
| $RF_{IMIN}$ | Reduction factor for instantaneous intrusion by the soil |
| $RF_{CR}$ | Reduction factor for time-dependent compression of the core (compressive creep) |
| $RF_{IN}$ | Reduction factor for time-dependent intrusion of the soil |
| $RF_{CD}$ | Reduction factor for chemical degradation |
| $RF_{PC}$ | Reduction factor for particulate clogging |
| $RF_{CC}$ | Reduction factor for chemical clogging |
| $RF_{BC}$ | Reduction factor for biological clogging |

All these effects are regarded as independent of one another, such that they can be multiplied to derive an overall reduction factor RF.

Note that the effects of weathering and installation damage are excluded. Since drainage materials will always be buried in the soil, weathering is not regarded as an issue provided that the index tests and time limits on exposure described in chapter 3.3 are fulfilled. Mechanical damage will have only an indirect effect on flow rate, but correct installation procedures are essential as with all geosynthetics.

$RF_{IMCO}$ and $RF_{IMIN}$ can be set equal to 1.0 provided that original method used to measure transmissivity uses the same level of compression and type of bedding as are anticipated in the soil. While this is by no means always the case, these tests can easily be performed in the short term.

$RF_{CR}$ and $RF_{IN}$ are measured as described in 3.5. No particular methods have been defined for the determination of the remaining factors. For chemical degradation use the methods described in 3.7.7.2 to predict the flow at the design life: $RF_{CD}$ is then the initial flow divided by this predicted flow. Biological degradation ($RF_{BC}$) can be caused by clogging and by root penetration for which tests such as EN 14416 exist, though not related to reduction in flow. No attempt is made to estimate the time over which roots of more familiar species might penetrate the nonwoven, attracted by the ready presence of water within the geosynthetic.

For drainage geosynthetics with a geonet core Giroud et al. proposed ranges of values for $RF_{CC}$ and $RF_{BC}$, together with similar ranges for $RF_{CR}$ and $RF_{IN}$, but were unable to propose values for $RF_{CD}$ and $RF_{PC}$ (table 3.13).

Table 3.13 Ranges of reduction factors for drainage geosynthetics with a geonet core after Giroud et al. (2000).

| Examples of application | Normal stress | Liquid | $RF_{IN}$ | $RF_{CR}$ | $RF_{CC}$ | $RF_{BC}$ |
|---|---|---|---|---|---|---|
| Landfill cover drainage layer Low retaining wall drainage | Low | Water | 1.0 to 1.2 | 1.1 to 1.4 | 1.0 to 1.2 | 1.2 to 1.5 |
| Embankment, dams, landslide repair High retaining wall drainage | High | Water | 1.0 to 1.2 | 1.4 to 2.0 | 1.0 to 1.2 | 1.2 to 1.5 |
| Landfill leachate collection layer Landfill leakage collection and detection layer Leachate pond leakage collection and detection layer | High | Leachate | 1.0 to 1.2 | 1.4 to 2.0 | 1.5 to 2.0 | 1.5 to 2.0 |

The procedure is therefore:
- Measure compressive creep at 20, 50, 100 and 200 KPa, with added shear when the application is a side slope.
- Predict the thickness at the design life.
- Establish a relation between thickness and flow to predict the flow at the design life. $RF_{CR}$ is the ratio of the initial flow to this predicted flow.
- Perform creep tests using hard bedding (no intrusion into the flow space) and using soft bedding. $RF_{IN}$ is the ratio of the first to the second. If this information is not available from creep tests, use the results from short-term tests under compressive loads.
- Alternatively use default factors for $RF_{CR}$ and $RF_{IN}$ from table 3.13.
- Use default factors for $RF_{CC}$ and $RF_{BC}$ from table 3.13.
- If chemical degradation is likely to obstruct flow (as opposed to causing complete collapse), estimate $RF_{CD}$.
- If particulate clogging is likely, estimate $RF_{PC}$.
- Then calculate the predicted flow at the end of the design life as the initial flow divided by $RF^{CR} \times RF_{IN} \times RF_{CC} \times RF_{BC} \times RF_{CD} \times RF_{PC}$.

290

- Separately, state the predicted time to collapse for a given compressive load and temperature or, conversely, the maximum compressive load that will not lead to collapse within the design life.
- Separately, state the time after which oxidation is predicted to lead to collapse or, conversely, state that collapse is not predicted to occur within the design life at the service temperature.

Do not allow for safety in the above factors as in addition there will be a factor of safety which will take into account structural factors. GSI(2001) and Giroud (2000)recommend a value of 2.0.

Note however that Müller does not advise the use of reduction factors to predict flow, but prefers determination of a design life under specified (required) conditions as described in chapter 3.8.

**Summary**
- For reinforcement applications:
  - Establish the characteristic tensile strength
  - Determine the reduction factors which represent the effects of sustained load ($RF_{CR}$), damage ($RF_{ID}$), weathering ($RF_W$) and the environment ($RF_{CH}$). Each factor refers to a particular design life and temperature and is derived from a specific series of tests. In some cases generic default factors have been proposed.
  - Determine the safety factor to take into account the uncertainties. Reduction and safety factors are combined in different ways.
  - Divide the characteristic strength by the reduction factors and the safety factor to give the design load.
  - Present creep strain as an isochronous diagram for a reference temperature. This is not subject to reduction factors. Any limit on strain can then be converted to a limit on design load.
  - For seismic design perform tests which establish the changes in material properties following a period of sustained or dynamic loading.
  - For oxidation (Mode 3) instead of calculating $RF_{CH}$ consider a simple comparison of predicted time to failure with the design life.
- For drainage geosynthetics:
  - Predict the time to collapse for a given compressive load and temperature or, conversely, the maximum compressive load that will not lead to collapse within the design life.
  - Predict the time after which oxidation is predicted to lead to collapse or, conversely, state that collapse is not predicted to occur within the design life at the service temperature.

Predict the flow at the end of the design life taking into account the effects of compressive creep, intrusion by soft bedding, chemical degradation of the core and particulate, chemical and biological clogging.

# References

*Chapter 3.7*

- Bathurst R J, Huang B, Allen T M. Analysis of installation damage tests for LFRD calibration of reinforced soil structures. Geotextiles and Geomembranes, Vol. 29, 2011, pp323-334.
- Bathurst R J, Huang B, Allen T M. Interpretation of laboratory creep testing for reliability-based analysis and load and resistance factor design (LFRD) calibration. Geosynthetics International, 19(1) in press (2012).
- Elias V. Corrosion/degradation of soil reinforcements for mechanically stabilized earth walls and reinforced slopes. US Federal Highway Administration Publication FHWA-SA-96-072, 1997.
- FGSV. Merkblatt über die Anwendung von Geokunststoffen im Erdbau des Straßenbaues, FGS Köln, 2005.
- GSI, standard GC-8, determination of allowable flow rate of a drainage composite.
- Giroud J P, Zornberg J G, Zhao A. Hydraulic design of geosynthetic and granular liquid collection layers. Geosynthetics International, 2000a, Vol. 7, pp 285-380. -
- Giroud J P, Zhao A, Richardson G N. Effect of thickness reduction on geosynthetic hydraulic transmissivity. Geosynthetics International, 2000b, Vol. 7, pp 433-452.
- Greenwood J H, Shen J M. Design loads and partial safety factors for geogrids in soil reinforcement. 5th International Conference on Geotextiles, Geomembranes and Related Products, Singapore. International Geotextile Society, 1994, pp 219-224.
- Hufenus R, Ruegger R, Flum D. Geosynthetics State of the art – Recent developments, eds Delmas P, Gourc J P. Proceedings of the Seventh International Conference on Geosynthetics, Nice, France. Balkema, Lisse, Netherlands, 2002, pp 1387-1390.
- Hufenus R, Rüegger R, Flum D, Sterba I J. Strength reduction factors due to installation damage of reinforcing geosynthetics. Geotextiles and Geomembranes, Vol. 23, 2005, pp401-424.
- Koerner R.M. Designing with geosynthetics, 4th edition, Prentice-Hall, 1997.
- Müller W, Jakob I, Tatzky-Gerth R. Long-term water flow capacity of geosynthetic drains and structural stability of their drain cores. Geosynthetics International, 2008, Vol. 15, pp437-451.
- Müller W, Jakob I, Li C S, Tatzky-Gerth R. Durability of polyolefin geosynthetic drains. Geosynthetics International, 2009, Vol. 16, pp28-42.
- Zornberg J G, Leshchinsky D. Comparison of international design criteria for geosynthetic-reinforced soil structures. Geosynthetics and Earth Reinforcement, ed. Ochiai H, Otani J and Miyata Y, 2001. International Society for Soil Mechanics and Geotechnical Engineering.

# 3.8    Conclusion

J.H. Greenwood

## 3.8.1    Assessment procedure

For **all applications** the procedure is as given in the following steps.

**Identify the geosynthetic.** The summary to chapter 2.1 lists the following characteristics:
*   Identification of the product (commercial name);
*   type of polymer (composition, morphology, additives, coatings);
*   physical structure of the geosynthetic: e.g. thick or thin fibres forming a woven or nonwoven fabric, extruded geogrid, coated fibrous strip, geosynthetic clay liner, continuous sheet;
*   joints forming part of the structure of the geosynthetic (e.g. woven, welded, integral);
*   mechanical and hydraulic properties (depending on function);
*   thickness or mass/unit area.

**Define its function and the environment.** Chapter 2.2 provides examples of the applications of geosynthetics and the environment likely to be experienced by the geosynthetic. In particular, define the temperature.

From the function, **identify the key property**, such a strength for a reinforcement, flow in the plane for a drainage geosynthetic, or permeability for a filter.

Define the **design life** (section 2.3.4) and the **design temperature**, and, if necessary, the concentration of aggressive media such as acids, alkalis and landfill leachates **in the design environment** (see sections 2.2.2 to 2.2.4). Where hydrolysis is a priority, define the local soil saturation.

From the above information, **identify and prioritise the most likely forms of degradation**. table 1.2 lists the most likely forms of degradation for particular functions. The examples of applications in chapter 2.2 also state the principal forms of degradation. As discussed in section 2.3.9.2, each form of degradation is considered separately, with the exception of temperature which in many cases will accelerate the rate of degradation.
The forms of degradation for consideration include:

*   Mechanical damage (see chapter 3.6).
*   Oxidation (see chapter 3.1).
*   Photo-oxidation due to UV (weathering) (see chapter 3.3).
*   Hydrolysis (see chapter 3.2).
*   Attack by acids, alkalis and landfill leachates (see chapter 3.1).
*   Leach out of additives (see chapter 3.1).
*   Compressive or tensile loading (see chapters 3.5 and 3.4).
*   Microbiological attack.

Forms of degradation that principally apply to thermoplastic geomembranes (geosynthetic barriers) are not listed. They include the swelling effect of solvents and environmental stress cracking. Freeze-thawing cycles, wet-dry cycles and ion exchange are effects specific to geosynthetic clay liners (clay geosynthetic barriers).

Decide whether assessment of durability is to take the form of:
- **Reduction factors** to be applied to the key property to give a reduced value that can be used in designDefinition of a **time-dependent property** as a table or graph to be used in design, rather than as a reduction factor.
- **Statement of lifetime**, for a single form of degradationUse of an **index test** to establish durability for a single form of degradation.

Examples of **reduction factors** are:
- For **reinforcement** applications durability will be taken into account by reduction factors which reduce the long-term strength of the reinforcement in design, together with predicted strains in the form of a table or a graph. This has been described in sections 3.7.1 to 3.7.6.
- For **drainage** applications durability will be taken into account by considering reduction factors for transmissivity (flow of water in the plane) due to compressive creep, and statements of lifetime for collapse together with chemical degradation of the core. This has been described in section 3.7.7.

Index tests are used for **applications with a design life of up 25 years**, subject to certain other conditions on the composition of the material and the environmental conditions, according to the European system. Durability is assured by the fact that the geosynthetic passes the index tests listed in section 1.7. As we have seen, equivalent tests are not yet available for durations of >25 years, as well as for the following situations:
- Where pH<4 or >9, in particular for polyester geosynthetics in highly alkaline environments with pH> 10, near concrete, lime or cement.
- When the soil temperature is > 250°C (also < 0°C, although degradation due to the mechanical effects of frozen soil is not considered).
- When recycled materials are used, provided that they are of a constant and controlled quality.

Other index tests, leading to appropriate **materials selection** include:
- Use of a robustness classification for mechanical damage (see section 3.6.3)
- Index tests, if acceptable, for the hydrolysis of polyesters (see chapter 3.2). According to ISO/TR 20432 these can be either fulfilment of EN 12447 (as for 25 years) or that the number-averaged molecular weight $M_n$, measured to ASTM D2857, exceeds 25.000 and that the carboxyl-end-group count is no greater than 30. In this case the material should be durable for 100 years in saturated natural soil at up to 25 °C with 4<pH<9, with some reduction in strength.
- Abrasion (see section 3.6.8)
- Protection efficiency (see section 3.6.9)

In all other situations, in particular for lifetimes greater than 25 years, **further testing** is required. Determine the method to be used, the property to be measured and the end-of-life criterion. This is discussed in chapter 2.3 and section 3.7.6. Examples are:

1. Oxidation testing as described in chapter 3.1. The oven ageing methods ISO 13438 A and B can be performed over a series of temperatures and times. Define a single end point, such a 50% retained strength. The tests should be planned as described in chapter 2.3 and the results used to create an Arrhenius diagram from which the lifetime under service conditions can be derived.

2. Perform the same procedure, and in addition monitor the stabiliser content by OIT or chemical analysis. Divide the duration of each test into two or three stages, characterised by
   - The anti-oxidant depletion time (stabiliser content unchanged)
   - The induction time to onset of polymer degradation (stabiliser content changes, but no reduction in strength)
   - The time to reach a drop of 50% in strength or elongation (ref. section 3.1.4.3 and figure 3.6 and 3.7).

3. Construct an Arrhenius diagram for each stage and extrapolate to find the duration of each stage under service conditions. Finally, add the durations for each stage to give the total service life.

4. Alternatively, use ISO 13438 Method C at different temperatures and oxygen pressures and extrapolate to the service temperature and an oxygen pressure of 0.21 bar, equivalent to the oxygen content of ambient air (see section 3.1).

5. Use a procedure as in a) to determine the lifetime in acids, alkalis or landfill leachates. If a higher concentration is used than is found in service, a test as in EN 14030 must be performed at different concentrations as in c) in order to establish the relation between concentration and lifetime. If leaching of additives has to be taken into account, divide the duration into two stages as in b), the first stage being the removal of additives and the second the subsequent reduction in strength.

6. For hydrolysis of polyesters use the procedure of EN 12447 can be extended to lower temperatures and longer times to yield an Arrhenius diagram (Schmidt et al. 1994). Use uncoated yarns. Chapter 3.2 describes the test methods and interpretation of the results in detail. Note that ISO/TR 20432 allows an index test for polyester yarns (see section 1.7.4).

7. The tests for the effect of mechanical loads in tension or compression can be accelerated as described in chapters 3.4 and 3.5. These methods are intended for reinforcement applications and may have to be adapted to suit the specifications for non-reinforcing applications.

For more information see Euro4 Keynote paper: "Long-Term performance and lifetime prediction of geosynthetics", by Hsuan et al (2008).

Long-term weathering presents a particular problem, since there is no method available to simulate long-term weathering within a reasonable time. This is discussed in chapter 3.3.

The same difficulty applies to microbiological attack. The test method in EN 12225 represents optimum conditions for this form of degradation, but these cannot be further accelerated. However, high molecular weight synthetic polymers commonly used in geosynthetics are in general not affected by the action of fungi and bacteria and the European system states that no testing is required on virgin (not recycled) polyethylene, polypropylene, polyester (polyethylene terephthalate: PET) and polyamides 6 and 6.6. The method must be applied to other materials including vegetable-based products, new materials, geocomposites, coated materials and any which are of doubtful quality.

**Predict the service life**. This should state:
- The geosynthetic.
- The environment for which the prediction is made:
  - soil gradation, angularity, compaction;
  - soil pH, presence of contaminants;
  - temperature;
  - saturation;
  - exposure to light.
- The property measured, and the end-of-life criterion.
- The predicted lifetime.
- The level of confidence.
- Any conditions applying to the prediction, e.g. correct installation practice.

Alternatively, **predict the properties of the geosynthetic at the design life** or express them as **one or more reduction factors**.

**State the consequences of the prediction**. For example state that, within the limits of current knowledge and assuming a certain environment, the geosynthetic will continue to fulfil its function throughout that design life, e.g. that material X, in environment Y, is expected to fulfil its function according to end-of-life criterion Z, for a minimum lifetime T. Note that some requirements may be higher at the installation stage (see section 2.3.3).
Alternatively, according to ISO/TS 13434, it can be stated that:
- The geosynthetic is sufficiently durable:
  - No change in the available property is predicted at the end of the design life.
  - A change in the available property is predicted at the end of the design life and the level is acceptable.
  - The ratio of the predicted available property to the predicted required property at the design life is acceptable.
  - The margin between the design life and the predicted end of life is acceptable.
- *The geosynthetic should be replaced* after a stated number of years. A *sample of geosynthetic should be extracted after a stated number of years* to determine the level of degradation; a decision regarding replacement will depend on the result.
- If replacement or extraction is impractical, the *geosynthetic is not sufficiently durable* for the application.

## 3.8.2 Expertise required to assess life time predictions

As shown in the previous chapters, durability testing is complicated and requires special equipment to perform. In Europe the laboratories accredited to do this testing are listed (see section 1.7.6). The evaluation of the results is even more complicated. Many factors and circumstances can influence the results and the extrapolation over decades is difficult to make.

The following notes are intended for the independent assessor who has to make, or more commonly check, a prediction.

- Make an unambiguous identification of the product. Demand a technical data sheet or website address and at least a small – typically A4 sized – specimen of each product covered (this provides physical proof that the product actually exists). Check its general appearance, particularly when the assessment is for a range of similar products.
- Check the test certificates. The reports or certificates describing the results of tests should state clearly the source and identification of the material tested as well as the test method and the results. Check that testing has been performed, or in exceptional cases witnessed, by independent or accredited laboratories.
- Check the strength and other requirements. The CE-marking certificate and quality assurance data may provide assurance of a characteristic strength, or more commonly a nominal strength. Check how this is defined. Check the results of control tests to see that the nominal strength is achieved – if not, question the data.
- Note the validity of index tests. Those in Annex B to the European geotextile applications standards EN 13249-13257 and 13265 assure 25 years' durability for the most common materials and environments. These are listed in chapter 1. Installation damage and creep are handled separately. The results of these tests are not sufficient on their own to ensure longer lifetimes, or to make predictions for more severe environments.
- Be conservative. If appropriate, add conditions on the manner in which the geosynthetic is stored and installed and in particular on the quality and inspection of joints. This is even more critical for geomembranes.

Particular problems arise when evidence is presented for similar but not identical products, or from exhumed material. The method for handling these is described in the following sections.

## 3.8.3 Use of data obtained on similar products

In many cases data is presented on a material that is similar to but not identical with that being assessed. A judgment has to be made on whether this data applies. In general, the two materials should be similar in fibre diameter and textile structure and consist of the same polymer with similar additives, with particular reference to the principal mode of degradation: additives for oxidation, CEG count and molecular weight for polyesters, tensile strength and modulus for mechanical properties. This should be confirmed by visual examination of samples as well as a limited programme of short-term tests, such as OIT testing for additives, tensile tests or one week creep tests for reinforcements. The results short-term tests can provide useful evidence on whether the products can be regarded as sufficiently similar. It is sometimes argued that newer grades of material should be better and that predictions based on earlier products should therefore be conservative, but remember that newer grades of material may have been introduced because they are cheaper, with a lowering of performance.

## 3.8.4   Data from exhumed material

### 3.8.4.1   Check list

At first sight, data from exhumed material forms an ideal basis for life prediction as it represents real life experience in a way that accelerated testing never can. There can be no questions about the material. However, the data refers only to the particular environment experienced: details of environment may not have been monitored and will it stay the same over the future life of the geosynthetic?

The following is a check list for data from exhumed material:

- What was the geosynthetic?
  - From the construction records or from labelling on the product determine the type of material used. Look for possible archive material, corners or cut-offs that may not have degraded – for example in comparing weathered material with unexposed material.
- Was it installed correctly, or is the degradation due to mechanical damage on installation?
  - Correct installation is an essential condition. Poorly installed geosynthetics will degrade and fail early due to the faults introduced. Changes in the strength of exhumed geosynthetics can often be attributed to installation damage alone, and that these changes mask any possible environmental degradation. Such results only stress the importance of exhuming some samples directly after installation, if only to provide a means of eliminating such effects when comparing results. Even then, the strength of damaged material is very variable and, in spite of eliminating the effects of the damage, the statistical variability of the corrected tensile strength may not allow any conclusion to be drawn about the presence or absence of environmental degradation. There have been several instances where any reduction in strength which might have been detected has been obscured by mechanical damage caused during the original installation.
- What was the environment?
  - What was the temperature pattern? If the material was exposed, what was the solar radiation and rainfall. If it was underground, where was it placed and what was the local saturation, soil granularity and pH? It may be necessary to refer to meteorological data on air temperature, rainfall or solar radiation from a nearby weather station.

In general the environment in practice will turn out to be more benign than that used in the original design.

- Were any measurements made on the material during service?
  - Ideally, one exhumation should have been made directly after installation, to allow the effects of installation damage to be measured and then eliminated, and a second exhumation halfway through the period over which the geosynthetic was in service.
- Was there a change in properties?
  - If so, were the measurements statistically significant?
- Can the measurements be extrapolated and how do they relate to failure or end-of-life?
- Are other forms of degradation present (e.g. weathering)?
  - Other forms of degradation must be isolated and, if they are not relevant to the conditions of lifetime prediction, eliminated by the use of control samples.
- What is the weakest point?
  - It may be necessary to allow for stress concentrations at joints or welded seams, for areas exposed to weathering, or to temporarily higher pH levels near concrete, for example.
- How does the time interval relate to the design lifetime?
  - Inevitably, most exhumations to date have been made after only a small fraction of anticipated design lifetimes.

The emphasis given to data from exhumed material depends on the appropriateness and precision of the data. Often there is too much uncertainty, for example the load applied is lower than was designed for, the soil temperature was not measured, the pH decreased with time, no archive material is available for comparison, or the exhumed material is not the same as today's product. Ultimately the assessor himself will have to decide on the quality of the data from exhumed material and the priority which it is to be given in the assessment.

### 3.8.4.2 Comparing data from exhumed material with the predictions from accelerated testing

One use of data from exhumed material is to check the predictions from accelerated testing. The results of accelerated testing should then be extrapolated to the conditions – in particular the temperature - experienced by the exhumed material, rather than the theoretical design conditions. Where a design temperature differs from the mean temperature experienced by the exhumed material, then the Arrhenius diagram may be used to compare the rates of degradation at these two temperatures and the design lifetime adjusted accordingly.

If the prediction equals the result of the tests on the exhumed material, then it provides a validation of that prediction. If the exhumed material performs better than predicted, then the prediction is conservative and the assessor may consider an adjustment. If the exhumed material performs worse than predicted, then it should first be checked whether the difference is due to damage while the geosynthetic was being installed, or when it was being excavated. If this is not the case, then it shows that the prediction is too optimistic and should be adjusted accordingly.

A problem arises when no change is recorded. There are many publications describing tests on geosynthetics that have been exhumed after a period in service with no conclusive evidence of degradation. Marketing managers delight in results demonstrating that "no degradation was observed", failing to mention that none was predicted for the limited time the geosynthetic was in the soil. All this tells the scientist that there is no unexpected premature degradation. Only if some change is actually detected and measured can the results be used for numerical lifetime prediction.

### 3.8.4.3    Extrapolation of the results from exhumed samples

The other possibility is to make a prediction directly from the results on the exhumed material. The geosynthetic should have been in service for at least a tenth of the design life, i.e. 10 years' service for 100 years' prediction, or longer if degradation takes place in one or more sequential stages. The measurements should be precise and sufficient in number for the rate of change to be detected and measured with statistical significance. The method of measurement should be identical throughout. Further details are given in ISO TR 13434 and ISO 13437. The change can then be extrapolated to the design lifetime. If the parameter measured does not change linearly with time, then the dependence will have to be deduced from the known behaviour of the material or empirically. There may be another property, such as stabiliser content, which can be measured more precisely or whose change precedes any change in strength. The general principles of extrapolation are given in chapter 2.3.

Müller-Rochholz et al (2002) made such a prediction for polypropylene fibres used in a landfill by monitoring the content of phosphate stabilizers over a period of 20 years. While a clear decrease was observed, it was necessary to assume an exponential relation, extrapolate it to 120 years, and assume a minimum stabilizer content of 10% as an end-of life criterion.

## 3.8.5    Planning for monitoring

While extrapolation of the results of regular monitoring provides an ideal, if site-specific, method for life prediction, where materials have been exhumed there is often a lack of critical information on, for example, the environment, or an absence of control material. This limits use of the results for other sites and applications. To provide for better information in the future, samples should ideally be installed with the deliberate intention of extracting them at set intervals in order to monitor the degradation. Although the evidence from installed samples will not be available for many years, well planned extractions will provide vital information, both in validating accelerated tests and possibly in warning of incipient loss of properties or of unexpected effects. Experience is the only definitive validation of durability and lifetime.

The following is a check list for the installation of samples for monitoring (see also ISO 13437):

Decide at the start on:

- the properties to be monitored;
- the frequency of measurement;
- the number and size of samples;
- indelible labelling of samples;
- the method of extraction, so as to minimize damage;
- how the environment is to be monitored;
- how and where the control material is to be stored;
- how the location of samples is to be recorded;
- how and where records are to be kept.

Note in particular:
- Choose properties that are likely to change within the timescale of the tests (e.g. intrinsic viscosity), even if these have then to be correlated with the functional properties (e.g. strength).
- Choose properties than can be measured with minimum variability.
- Choose properties that can be compared with the results of accelerated testing.
- If possible set up long-term laboratory tests on the material at temperatures that are elevated but require extensive time periods, for example 60, 70 and 80°C, and monitor the rates of degradation.
- Include reserve samples.
- Choose neighbouring samples for exposure and control purposes
- Measure temperature, if possible pH and soil moisture content, and for exposed samples the radiant exposure.
- If high surface temperatures are expected, install samples close to the surface, since higher temperatures increase the rates of ageing and creep of polymers disproportionately.
- Ensure that the method of extraction should minimize disruption to the structure and damage to the samples.
- Eliminate the effect of mechanical damage by installing and then immediately extracting samples which will then have been subject to installation damage but not to long-term degradation.
- A paper copy of all records should be kept in addition to electronic data in to ensure that they can be retrieved in many years' time.
- A cool and dark environment is recommended for the control material; avoid PVC-bags because of softener-migration.
- When these are established the durability can be reassessed with a higher level of confidence.
- Do not expect immediate results: be patient.

### 3.8.6    Remanent life assessment

As geosynthetics approach the end of their design life the question will undoubtedly be asked: will they last another, say, fifty years? The more expensive they are to excavate and replace, the more important the question. Surely, it may be argued, the environment experienced by the geosynthetic has been more benign that was originally allowed for in design? There must be life left in the old dog.

In other fields of industry a considerable science has evolved in remanent life assessment, particularly in power stations and petrochemical plant. They have the advantage in that the operating conditions of such a plant – unlike a landfill or a reinforced wall – are generally well documented. The first stage in such an assessment is therefore to go through the original design process inserting the actual operating conditions – in our case environmental conditions – instead of those used in the original design. This will give an initial estimate of how far degradation should have proceeded – how much creep strain, how much remaining stabilizer. The next stage, and a more expensive one, is to perform a physical examination – to measure and observe on site, and to take samples for examination in the laboratory. These measurements should be made such that they can be compared directly with the calculated predictions, and they can be used to modify those predictions. Real experience always dominates over theory.

Having achieved a uniform picture of the rate of degradation, conclusions can be drawn. The simplest and most desirable conclusion is that the geosynthetic will continue to function for the required life extension. The undesired conclusion is that it has already passed its end-of-life criterion and requires immediate replacement. In between the two are conclusions such as to reassess the material after a set time, or to introduce regular monitoring with a predetermined limit.

Such procedures are for the future, but note that there is already an established methodology which has proved itself in other branches of industry.

### 3.8.7    Final remarks

> Former US Defense Secretary Donald Rumsfeld's quotation. Much mocked at the time, it is in fact a perfectly logical statement that applies equally to life prediction in geosynthetics.
>
> *...as we know, there are known knowns; there are things we know we know. We also know there are known unknowns; that is to say we know there are some things we do not know. But there are also unknown unknowns - the ones we don't know we don't know.*

So far the durability of geosynthetics has proved to be very good. The most common causes of degradation have been incorrect materials selection or structural design, followed by installation damage and weathering. Very few changes have been attributed to oxidation or hydrolysis. With geomembrane failures caused by poor material selection and incorrect installation have led to improved materials, better installation practice and quality control. Never forget that life prediction is only of any value if the geosynthetics are installed correctly.

Do not forget how long 100 years is. This document is being prepared in 2012. 1912 saw the sinking of the Titanic, Amundsen's return from the South Pole and the Stockholm Olympics. Much has happened since.

Any prediction is an estimate. Nobody expects that on completion of its predicted lifetime a geosynthetic will lose all its strength, any more than bread bought from a food market is found covered in grey mould the day after its printed sell-by date. Degradation is slow, fickle, and sensitive to the environment. A prediction can only take into account the known forms of degradation – the 'known knowns' in Donald Rumsfeld's famous quotation reproduced in the box above. A safety factor may cover the 'known unknowns'. But the prediction can never take into account the wholly unexpected – the 'unknown unknowns'. One of the authors scared a geosynthetics manufacturer by suggesting that a biological organism could mutate into a form that would devour his geogrid – and then spent an hour explaining that this was science fiction. Climate change will change the environment, but we are not sure how. The prediction is the best we can do with what we know, and those that follow us will either praise or disown our efforts long after we have joined the company of geosynthetics deep under the soil.

**Summary**
- A procedure is set out for determining the durability of geosynthetics (except where reduction factors are used).
- From this it may be concluded that the geosynthetic is or is not acceptable for use, or that it should be inspected and/or replaced at set intervals.
- Advice is given on the items to be checked, the use of data from similar products.
- Data from exhumed material may be taken into consideration under certain conditions. It can be used to validate the predictions from accelerated testing.
- Guidelines are given for the monitoring of geosynthetic structures, where this is possible.
- Guidelines are given for the assessment of the remaining life of a geosynthetic that has been in service for some years.
- So far the durability of geosynthetics has proved to be very good: most failures have been due to incorrect materials selection or faulty installation, emphasising again the importance of correct installation procedures for durability.
- Any prediction is an estimate only.

# References

*Chapter 3.8*

- Hsuan Y G, Schroeder H F, Rowe K, Müller W, Greenwood J H, Cazzuffi D, Koerner R M. Long-term performance and lifetime prediction of geosynthetics, 2008. EuroGeo4, Edinburgh, UK, pp 327-366.
- Müller-Rochholz J, Bronstein Z, Schröder H F, Ryback Th, Zeynalov E B, von Maubeuge K P. Long-term behaviour of geomembrane drains – excavations on landfills after up to 12 years service. Geosynthetics – State of the art – Recent developments, eds Delmas P, Gourc J P. Proceedings of the Seventh International Conference on Geosynthetics, Nice, France. Balkema, Lisse, Netherlands, 2002, pp565-568.
- Schmidt H M, Te Pas F W T, Risseeuw P, Voskamp W.; The hydrolytic stability of PET yarns under medium alkaline conditions; 5th International Conference on Geotextiles, Geomembranes and Related Products, Singapore, 1994. International Geotextile Society, 1994, pp 1153-1158.

# CONTENTS

# CHAPTER 4
# GEOMEMBRANES

J.H. Greenwood

## 4.1    Introduction

Geomembranes are used as barriers that are largely impermeable to moisture and most liquids. In common with the rest of this Guide, this chapter is intended to cover their durability when used in building, construction and normal soils. Geomembranes are however widely used elsewhere in civil engineering, such as in tunnels, dams, reservoirs and canals and in ponds and landfills for solid and liquid waste, where the liquid or leachate can be chemically aggressive and where leakage could lead to contamination of the underlying soil and of the groundwater. Since the durability of geomembranes has been studied particularly with these applications in mind, and corresponding tests developed, one cannot review the subject without reference to them. For readers specifically interested in landfill applications a summary is given below and further references are given.

Fig. 4.1 Reservoir with geomembranes (Courtesy GSE Lining Technology GmbH)

High (or more correctly medium) density polyethylene, described in Section 2, is a common choice for geomembranes used as liners thanks to its price, general durability and the relative ease of welding. The success rate with these materials has been good, and the measures necessary to achieve this will be discussed. Related materials such as linear low density polyethylene (LLDPE), chlorinated polyethylene (CPE), chlorinated sulfonated polyethylene (CSPE) and polypropylene (PP) are not covered in this chapter pending more evidence on their durability.

PVC has been used for the protection of concrete and the lining of tunnels, reservoirs, dams and canals. PVC's durability is described below. Geosynthetic clay liners are summarized.

307

Bituminous liners have been used on dams, in tailings ponds, and other applications in cold environments, but, are not covered by this chapter, but a review given by Yu et al. 2011 is added to the list of references for the ageing of polymer modified bitumen..

In most of these applications repair and replacement are difficult or impossible and we shall there-fore assume that a design lifetime of at least 100 years is required.

## 4.2    HDPE

### 4.2.1.  Six stages of degradation

The long term performance of geomembranes depends critically, far more critically than for geo-textiles, on the care taken during design, materials selection and above all on best practice installation. While faulty construction practices and covering with backfill or waste can create holes in the liner directly, stresses left over from folds or welding provide initiation sites for stress cracking with or without the help of oxidation. Needham et al (2004b) derived a conceptual model comprising the following six stages of degradation in geomembranes, based on their application in landfills:

1. Holes arising before or during construction of the liner and placement of the cover and drainage system. An electrical leakage survey allows these holes to be identified and repaired.
2. Holes arising during backfilling by mechanical damage or the extension of existing defects. Again, an integrity survey will identify these and it should be possible to repair them.
3. A dormant period during which no further holes are expected to develop.
4. If the membrane is locally stressed, the development of stress cracks, before the onset of oxidation. This will depend on the magnitude of the stresses and the temperature.
5. Further stress cracking, this time assisted by embrittlement of the material due to oxidation.
6. Slow extension of existing holes and generation of new ones. The geomembrane leaks at the holes but remains intact elsewhere.

### 4.2.2  Stages 1 and 2: Mechanical damage

Direct mechanical damage in geomembrane liners occurs during installation and seaming (Stage 1), or when backfill or waste is deposited (Stage 2). The extent of the damage can be identified using mobile leak detection methods after seaming, which allows for immediate repair, or using built-in leak detection systems after filling, at which point leaks can be repaired if the geomembrane is uncovered. Nosko (1996) identified 24% of defects as having arisen during installation of the liner, 73% as occurring during placement of the cover soils, and 2% due to accidental damage by vehicles or the installation of pipes and drainage systems. Other data suggest a higher proportion of damage caused by vehicles after construction. Durability depends on the correct choice of material together with protection and installation procedures to limit mechanical damage, backed up by leak detection which allows for larger defects to be identified and sealed before installation is taken any further. Leak detection is more commonly applied in landfills where a double liner may be required: this is not currently the situation in Europe. Needham et al conclude that:

• Unless detected and repaired, holes may be present in liners due to imperfect manufacture and installation.

• Roughly two-thirds of holes occur during installation and one-third during backfilling and subsequent operation

• A fixed leakage detection system will enable hole detection location and in most cases will facilitate repair.

• Defects are most likely to occur at exposed or poorly protected liner at the margins, on bunds, benches and slopes.

The emphatic conclusion for life prediction is that unless the membranes are correctly installed and welded, failure will occur earlier than predicted on the basis of material properties alone.

Fig. 4.2 Installation of geomembrane (Courtesy GSE Lining Technology GmbH)

### 4.2.3   Stage 3: dormant period

Monitoring of a landfill site in the UK showed that holes were rarely found to develop during the first eight years of service as a result of stresses imposed by the waste alone, and those that did occurred relatively quickly. Evidence from the USA indicates that significant numbers of holes do not develop as a result of degradation, stress cracking and ductile failures in the first ten years of operation. Needham et al therefore conclude that further holes are unlikely to develop during the first ten years of operation of the landfill (Stage 3).

### 4.2.4    Stage 4: stress cracking

When a specimen of HDPE geomembrane is stretched at a constant rate, as in a tensile test, the load reaches a yield point, after which the load slowly decreases while the material draws out. Final 'ductile' failure occurs at an extension of several hundred percent. A similar, ductile failure occurs under a constant load when this load is very high, close to the yield point. A relation can be established between load and time to failure, typically

$\sigma = t^n$

where $t$ is the time to rupture, $\sigma$ is the applied load and n is a number of the order of 20-40. If how-ever an intermediate load is applied and held constant, after a certain time the specimen will be found to break after a shorter time than predicted and, notably, at a much lower extension. The fracture surface is flatter and smoother than the contorted "fibrillar" surface typical of ductile failure, so that this form of failure is referred to as 'brittle', although not in the manner of shattered glass.

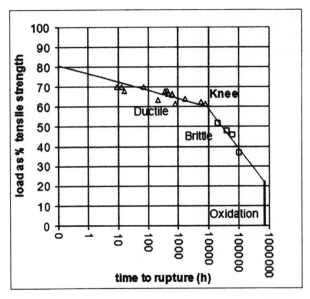

Fig. 4.3 Creep-rupture (stress-rupture) diagram for HDPE geomembranes.

The relation between load and time to failure for these two types of breaks is shown in Fig. 4.3. The transition from ductile to brittle failure leads to a change in gradient, generally referred to as a "knee". Above the knee ductile failure takes place at short times before brittle failure can develop. The slope of the graph is low. Below the knee brittle failure occurs first and the slope is steeper.

'Brittle' failure in HDPE and other semicrystalline polymers is a complex process. The crystallites in the semicrystalline mix are connected across the intervening randomly structured (amorphous) regions by looped and tie molecules which carry the mechanical load applied from outside. In time some of these tie molecules untangle themselves from the adjoining crystallites, a process which is thermally activated and which therefore obeys Arrhenius' formula. With the disentanglement and rupture of some tie molecules the load is shifted on to their neighbours, which in due course will also break.

Cracks start typically at small defects in the material such as catalyst residues and carbon black ag-glomerates, or at points of stress concentration such as sharp internal angles. At first the infant crack will be crossed by fibrils of oriented material, similar to the tie molecules but larger. These incomplete cracks or 'crazes' can sometimes be seen as a reflective shimmer in transparent materials such as old polymethyl methacrylate ("Perspex" or "Plexiglas"). Crazes grow slowly into fully open cracks which propagate along the interface between the crystallites, resulting in a relatively smooth, undeformed fracture face. Conversely, if the load is higher, or applied more quickly, as in a conventional tensile test, the loops and tie molecules do not disentangle from the adjacent crystallite but remain embedded in it, causing it to break into small segments that manifest themselves as material stretching.

'Brittle' fracture in HDPE is extremely well charted and understood because it limits the pressure rating of HDPE gas and water pipes, where failure can have consequences even more immediate than in geomembranes. To the annoyance of quality assurance managers looking for instant answers, it cannot be measured over short periods because above the knee ductile failure intervenes. Longer tests under sustained load (or pressure) must be performed. However, since crack growth is thermally activated fracture can be accelerated by raising the temperature, shifting the two branches of the creep-rupture characteristic to shorter times. Shorter term pressure tests at high temperatures can therefore be used as proof tests to ensure sufficient resistance to creep-rupture at service temperatures - provided, that is, that ductile failure is not allowed to intervene. Such tests are performed routinely on extruded polyethylene pipes.

Polymers with long molecules and a high density of side chains have proved more resistant to this form of cracking. Highly oriented polymers are also less susceptible, which is why this 'brittle' failure is generally ignored in highly oriented geotextiles, though it has been reported for polypropylene fibres (Seeger et al. 2002) and at the nodes of early extruded polyethylene geogrids (Peggs and Kanninen 1995, Bright 1996). Although this has been corrected by careful materials selection, the possibility of 'brittle' failure in extruded geogrids still haunts reinforcement design and has led to correspondingly conservative safety factors.

In geomembranes, unlike gas pipes, the stresses in service are not accurately known. It is evident that these stresses can be caused by incorrect placement leaving behind wrinkles, folds and indenta-tions, by not allowing for thermal expansion and contraction, by trafficking, by down-drag on the sides of landfill and locally by solid sharp objects. Without an accurate knowledge of these stresses numerical prediction of lifetime is impossible. Stresses have to be minimised by correct installation and quality assurance and the times to failure in practice will only be established by regular leakage monitoring, where this is required, and

311

by experience. Peggs and Elie (2008) describe the complex environmental factors that can influence stress cracking.

One mitigating factor is stress relaxation. Residual stresses left behind after installation will diminish with time thanks to stress relaxation, which is the counterpart of creep as described in Section 3.4.4. This process, which occurs faster at higher temperatures, will reduce the likelihood of failure. A further effect is that of physical ageing. Unlike in metals, the structure of a polymer is not static as soon as it is cooled below its crystalline melting point. Instead, there is a slow reordering and stabilization of the molecular structure. Stress relaxation, which occurs when the material is static, and creep which occurs when the material is allowed to strain under an applied load, both proceed faster in a young geomembrane than in an older one. As the geomembrane ages, or more precisely as its molecular structure stabilizes, the processes of creep and relaxation proceed more slowly. This is known as physical ageing. Since geomembranes are generally stored for weeks or months before installation, physical ageing is fairly complete for the bulk sheets, but it can be significant for welded material which is briefly remelted and often placed under load soon afterwards. However, Soong and Koerner (1997) have shown that complete stress relaxation does not occur, typically not getting below 22% of the initial stress. It is therefore important that welded material be allowed to stabilise before being placed under load.

*Environmental stress cracking*

Brittle failure of HDPE geomembranes, and for thermoplastics in general, can be greatly accelerated by certain, generally organic, chemicals. This form of failure, known as Environmental Stress Cracking or ESC, is the most common form of unexpected failure generally in plastics, partly because it has only recently been fully understood. On first sight it appears to be similar to stress corrosion cracking in metals, in which mechanical stress in the presence of an aggressive chemical can lead to more rapid failure than if the stress was absent, the bonds between neighbouring atoms being more vulnerable to chemical attack when under stress. ESC, however, is more a physical than a chemical effect. The liquid must be capable of diffusing through, dissolving in and being absorbed by the polymer, where it causes swelling and weakens the bonds between the polymer chains. The 'tie' molecules that would have resisted growth of the developing crack are more easily pulled out of the swollen polymer: consider them lubricated. The whole pattern is similar to cold spaghetti; on its own it forms a sticky mass from which it is almost impossible to extract a single piece of spaghetti (read: polymer chain) unbroken, but if tomato sauce or olive oil (read: organic solvent) is added before the spaghetti cools, then after cooling the individual pieces of spaghetti can be unravelled without difficulty. Thus the liquids most likely to cause ESC are not aggressive chemicals but organic solvents that dissolve in and swell the polymer, such as aromatic hydrocarbons, and those that reduce surface tension, such as detergents. These solvents are found very widely, such as in many cleaning fluids, adhesives and foodstuffs and as additives to concrete. ESC fractures are characterised by their very smooth surfaces up to the point where ductile failure takes over.

*Index test for stress crack resistance*

Hsuan (1998) tested HDPE geomembranes in tension and recorded the times to failure at different temperatures. She found that provided the temperature was raised no higher than 70 °C the times to failure follow an Arrhenius relation, such that failure times at 70 °C are less than those at 50 °C by a factor of 4.7. This demonstrates that one can predict failure lifetime for ESC if the environment, temperature and mechanical stress are known.

While the environment and temperature in service may be known, the mechanical stresses in service are not known precisely (ideally they should of course be zero). Instead, sufficient resistance to stress cracking is assured by means of an index test in which notched specimens are suspended under load in a particular surfactant (EN 14576; ASTM D-5397). Failure of the notched specimen in the surfactant may occur in not less than 200 h. This is regarded as sufficient to ensure a long lifetime for HDPE geomembranes, even for use in landfills, although longer test durations have been specified. Peggs (2010) describes the subsequent cracking of a HDPE geomembrane which passed the tests current at the time of its manufacture. The stress cracking of PP is described by Peggs (2008).

*Strain limit*

If the index test is to ensure sufficient resistance to failure by stress cracking it must be accompanied by measures to limit mechanical stresses during installation, for example by specifying suitable protection, avoiding wrinkles and folds, applying correct welding procedures, and considering alter-native membranes for the storage or containment of organic chemicals or hydrocarbons.

The use of strain limits to demarcate polymer behaviour has long been promoted in Germany in particular. A global limit of 3% strain in an HDPE geomembrane can be regarded as a good estimate for a long-term strain limit to prevent the risk of stress cracking (Seeger and Müller 2003). Lower limits have been proposed for local stone protrusion and for welded seams.

## 4.2.5  Stages 5 and 6: oxidation

The final phases in the degradation of HDPE and other polyolefins concern oxidation. The chemistry of oxidation and the functions of the various antioxidant additives have been described in Section 3.1. Once the antioxidants have been depleted by migration, leaching or consumption to the point when they are ineffective (the incubation period), oxidation sets in throughout the polymer. However, it should be noted that oxidation first occurs on the surface of the geomembrane then progresses to the centre. Oxidation attacks the polymer chains, reduces the molecular weight, embrittles the geomembrane and renders the material more sensitive to stress cracking. If the duration of oxidative attack is relatively short compared with its incubation period of its initiation, it appears on a failure diagram as a vertical cut-off (Fig. G1). Oxidation is the ultimate factor limiting the life of ge-omembranes.

The rate of oxidation and thus the lifetime depend critically on the local temperature, as mentioned in the next section. In water or aqueous solvents migration and leaching of the antioxidants can increase the rate of degradation, particularly when combined with oxidation (Hsuan and Koerner 1998, Sangam and Rowe 2002, Müller and Jakob 2003).

## 4.2.6    Landfills

The presence of transition metals in solution will further promote oxidation. In landfills the com-position of a leachate from municipal solid waste is variable but consists principally of volatile fatty acids, inorganic salts, surfactants and trace transition metals. These, particularly the surfactants, can accelerate degradation and stress cracking, thus immersion tests in simulated and regularly refreshed leachates are necessary to make a valid life prediction. The frequency of replacement of the leachate and the overall duration of the test depend upon the detailed chemistry: an understanding of this is essential if the test is to be meaningful. This is a complex area and the reader is referred in particular to papers by Rowe and his colleagues. Rowe et al. (2008) derived an activation energy of 62.5-64 kJ/mol, found that antioxidant depletion was faster in acid (pH4) and alkaline (pH10) environments, and recommended specific test leachates. Rowe and Rimal (2008) indicated that the antioxidant depletion rate in a leachate containing surfactant is 2.5 to 4.0 times faster than in water and 2.8 to 6.7 faster than that in air. An Arrhenius diagram predicts a lifetime in leachate of just 25 years at 20 °C, and 5 years at 50 °C. In air the corresponding values are 190 and 25 years. Use of a leachate-soil mix is less aggressive than immersion in liquid leachate, as is the use of a sand or other protective layer in practice. Analysis of Stages 2 and 3 of the three-stage oxidation model (see Chapter 3.1) lead to overall service lives of over 1000 years at 20 °C reducing to 65-90 years at 50 °C.

Fig. 4.4 Landfill cover (Courtesy GSE Lining Technology GmbH)

## 4.2.7    Durability tests for geomembranes

Chapter 3.1 describes the index tests for oxidation to assure sufficient life for HDPE geomem-branes, and the use of similar methods to make a specific life prediction. The test methods and the scientific bases for the tests have been described in detail by Needham et al. (2004a, 2004b), Rowe (2005) and Müller (2007). Annex B common to EN 13361, EN 13362, EN 13491, EN 13492, EN 13493, and EN 15382, specifies the additional tests required to assure the durability of geomembranes in Europe without reference to any particular lifetime: these include tests for weathering during installation (EN 12224), microbiological degradation (EN

12225), ESC (EN 14576), leaching by water or aqueous liquids (EN 14415) and chemical degradation by leachates (EN 14416). A comprehensive approvals process is operated by BAM in Germany, but note that the approvals authority is given access to the formulation of the polymer used, information not generally available to other users.

## 4.3   PVC

Polyvinyl chloride (PVC) geomembranes have been used worldwide in the rehabilitation of old concrete and masonry dams, in reservoirs and canals and providing floating covers for potable water reservoirs. This section is based on the reviews by Hsuan et al (2008) and Cazzuffi et al. (2010).

Plasticized PVC contains approximately 60% of basic PVC resin, characterized by its K value or molecular weight and viscosity. A high molecular weight improves the mechanical properties but makes the material more difficult to weld. 25 to 35% comprises plasticizers to lower the glass transition temperature and increase flexibility. The remaining constituents include:

- Lubricants to improve processing.
- Pigments, including carbon black and white titanium dioxide.
- Stabilizers against heat and ultraviolet light.
- Fillers.
- Additives.
- Reinforcements where mechanical strength is needed.

Depending on the local specification, PVC may also contain a proportion of recycled material. Note that there are large differences in the quality of PVC used in Europe and in North America.

If plasticizers are the key to making a PVC geomembrane flexible enough for use, then loss of plasticizer is the key to its degradation. Typically, a new membrane fails at 500% elongation, while an aged membrane yields and then fails at about 130% elongation. The density increases, the volume shrinks, and with the lower elongation at break this can lead to local cracking following an impact, especially at welds. At low temperatures the glass transition temperature increases, typically from -30°C to 0°C, increasing the probability of failure on cold nights.

If the geomembrane is exposed to air (with or without light) the plasticizer can be lost by evaporation, depending on its molecular weight and formulation, while if it is in contact with a liquid the plasticizer may dissolve, be leached from the surface, be hydrolysed or be subject to biological attack. A low molecular weight liquid can even penetrate the PVC and extract the plasticizer. The plasticizer can also migrate across a boundary into another polymer. Local stresses, such as occur on a rough sub-base, increase the rate of loss. This rate becomes slower with time as the plasticizer content decreases.

Higher lifetimes have been achieved by the correct choice and quantity of plasticizer, the key properties being its molecular weight and morphology (chain branching). A molecular weight of at least 400, greater than that used in the early materials, effectively eliminates the three loss mechanisms mentioned. A variety of different plasticizers are used, notably in Europe diisodecyl phthalates. Solid polymeric plasticizers have been employed in PVC membranes used to line Alpine dams without any evidence of degradation after 40 years.

PVC geosynthetic barriers have a high chemical resistance to the majority of acids, bases, salts and alcohols, but both the plasticizers and the PVC itself can be affected by organic solvents, even in very small concentrations. This is of less concern in general civil engineering, though the solvents may be present in landfills. Exposure to aqueous and especially alkaline solvents may however accelerate ex-traction, for example in tunnels.

PVC itself degrades chemically by dehydrochlorination and oxidation, principally when exposed to UV light or more slowly in the presence of heat. This is a chain reaction in which double bonds are formed in the polymer chain, releasing HCl – thus the name dehydro-chlorination. These double bonds absorb visible light, causing discoloration, and can also react with oxygen producing carbonyl products, leading to chain scission, crosslinking and a reduction in mechanical properties. Stabilizers may be added to suppress these reactions.

The plasticizers themselves can degrade. In some PVC materials the UV light directly affects the plasticizer causing rapid and premature failure. Others are sensitive to alkaline environments.

A variety of different plasticizers are used, notably diisodecyl phthalates. Solid plasticizers have been used in PVC geomembranes used to line Alpine dams without any evidence of degradation after 30 years (Cazzuffi, 2013), see also Fig. G.2.

Generally, lifetime prediction of PVC depends on accurate assessment and extrapolation of the loss of plasticizer. It is generally accepted that a minimum of 20% plasticizer is required for the geomembrane to function properly. Thus if the initial content is 33% and the end point is 20%, the lifetime corresponds to a 40% loss of the plasticizer present.

Where time and location permit, samples should be taken from the geomembrane at regular intervals and the plasticizer content or elongation at break plotted against time (Cazzuffi, 1998). Some graphs are relatively linear, making extrapolation to an agreed end point such as 20% plasticizer content simple. If this process is too slow, further samples can be aged at elevated temperatures in order to accelerate plasticizer loss and Arrhenius' formula applied (Hsuan et al 2008, Shimaoka et al 2010). As with HDPE, incorrectly prepared seams can lead to regions of weakness and possible potential failure.

The lifetime of correctly stabilized PVC geomembranes is generally not less than 30 years and in some cases has been predicted to be much longer. Since the rate of loss depends strongly on the type of plasticizer used as well as on the environment, it is difficult to transfer predictions from one PVC geomembrane to another.

## 4.4    Geosynthetic clay liners

Geosynthetic clay liners (GCLs) or clay geosynthetic barriers (GBR-C) consist of a layer of sodium or calcium bentonite clay sandwiched between two woven and/or nonwoven geotextiles. The sheets are held together by needlepunching or stitching, and the bentonite needs to hydrate under confining stress to be effective. In some products the needled fibers are heat burnished on one side to improve the mechanical bonding. Generally, an adequate soil pressure minimum 500 mm should be maintained to ensure effectiveness Adjacent panels of GCL are overlapped rather than been mechanically seamed together. The barriers were first used in their present form in the USA in 1985 to control leakage at a double lined Illinois landfill. They are often used under a geomembrane as a composite liner. In 2002 the United States Environmental Protection Agency (EPA) published a report summarizing the leakage records of 279 landfills, all double lined with leakage protection, with ages between 5 to 15 years. The leakage rate was found to decrease consistently to effectively zero. GBR-C are now widely used in landfill capping.

Lifetime prediction of these barriers is still in its infancy. This chapter is concerned with the long-term durability of these barriers and not with the possibility of short-term failure due to desiccation of the clay or cycles of freeze and thaw. They are also susceptible to damage during installation, and pulling apart of panels due to inadequate overlaps, particularly on slopes, is a common mode of failure. Possible causes of long-term failure are seen as:
- Loss of strength of the geotextile to the chemical environment, particularly leachates, and including oxidation
- Non-uniform confining pressure
- Long-term rupture under shear forces
- Penetration by roots
- Change in permeability of the clay following ion exchange
- Desiccation of the clay by cycles of wet and dry or freeze and thaw, or by water extraction by roots

Life prediction of a nonwoven polypropylene geotextile is described in Chapter 3.1: lifetimes of over 100 years have been predicted. Long-term shear is important in applications of GCLs where, for example, the barrier is used to line the sloping walls of a landfill. Failure is caused by the disentan-glement and rupture of the fibres that join the two layers of nonwoven goetextile, whether nee-dlepunched or stitched. The time to failure can be predicted by performing creep-rupture tests under identical shear loads and a fixed normal load at different temperatures. Müller and Seeger (2004) measured the time to failure of GCLs in shear, with particular reference to the sides of landfills, and established an Arrhenius relation relating time to failure with temperature. They confirmed that high short-term strengths do not necessarily indicate a long time to failure, and that long-term testing is essential. The nature of the water used has a significant effect.

Resistance to roots has been the subject of a proposed index test using lupins, which proved un-satisfactory (CEN/TS 14416). Roofing membranes are tested using a pyrocantha with a powerful tap root, but the test takes two years (EN 13948). Current practice is to rely on the mechanical properties of the nonwoven layers.

Ion exchange can take place in GCLs after they have been installed, and consists of a replacement of sodium ions by calcium to produce calcium bentonite. While tests have monitored this change, the timescale over which it occurs, and any changes in permittivity, they have not yet been extended to long-term life prediction. Cycles of wet and dry, and freeze and thaw, have been tested and modelled.

In conclusion, the lifetime of GCLs could be limited by a number of factors of which attempts have only been made to predict chemical degradation of the geotextile and failure in shear. As with other geosynthetic liners, selection of the right material and correct installation are essential.

**Summary**
- Geomembranes are used as barriers to liquids in building and construction. They are also used extensively in landfills, liquid containment, tunnels, canals, dams and reservoirs.
- HDPE and other polyolefins are the most common geomembrane materials used. Their success rate has been high, given the essential conditions of good design, correct materials selection and accreditation and above all the best practice in installation and quality assurance. All these result in a good underground lifetime.
- Geomembranes, particularly HDPE, should function solely as a barrier and should not be load-bearing members of lining systems.
- Holes arising during placement or waste filling can be detected by a leakage survey and repaired.
- Mechanical stresses, due to point loads, folds or welds, can over time lead to cracking of the membrane. These cracks can grow into holes or tears. Certain chemicals, such as organic solvents, can greatly accelerate this process. The index test for stress cracking is used to screen materials which are more susceptible to this failure mode, coupled with a strain limit in design.
- Ultimately the lifetime of HDPE and other polyolefin geomembranes is likely to be set by their resistance to oxidation, as described in Chapter 3.1. The rate of oxidation depends critically on the local temperature and can vary widely, due to the great variety of possible blends, copolymers, additives, morphologies and designs. A full assessment can only be made if the formulation is known; screening tests are specified where this is not the case.
- The use of HDPE geomembranes in landfills is complicated by the chemistry of the various leachates and the high local temperatures and is only summarized here.
- The lifetime of PVC geomembranes is governed by the loss of plasticizer. Where samples are taken and investigated, the loss of stabilizer can be plotted against time and extrapolated to an end point such as 20% plasticizer content.
- The lifetime of correctly stabilized PVC geomembranes is generally not less than 30 years and in some cases has been predicted to be much longer. Since the rate of loss depends strongly on the type of plasticizer used as well as on the environment, it is difficult to transfer predictions from one PVC geomembrane to another.
- The lifetime of a geosynthetic clay liner (GCL) can be limited by degradation of the polypro-pylene geotextile, failure in shear or by pulling apart of overlaps, root penetration and various factors affecting the clay. Failure of the polypropylene is described in Chapter 3.1 and failure in shear can be predicted by accelerated testing. As with other geosynthetic liners, selection of the right material and correct installation are essential.
- For further reading see in particular the publications by Cazzuffi, Müller, Hsuan, Peggs and Rowe and their colleagues, of which a selection are given below.

# References

- Bright, D. HDPE Geosynthetics: Premature Failures and their Prediction: Discussion and Closure. Geosynthetics International, 1996, Vol. 3, No. 1, pp 145-151.
- Cazzuffi, D. Long term performance of exposed geomembranes on dams in Italian Alps, Proceedings of the 6th International Conference on Geosynthetics, Vol. 2, Atlanta, USA, 1998, pp. 1107-1110.
- Cazzuffi, D. Long-time behaviour of exposed geomembranes used for the upstream face rehabilitation of concrete and masonry dams", Proceedings of the 9th ICOLD European Club Symposium, Venezia, Italy, 2013, p. 8.
- Cazzuffi, D., Giroud, J.P., Scuero, A., Vaschetti, G. Geosynthetic barriers systems for dams, Keynote Lecture, Proceedings of the 9th International Conference on Geosynthetics, Guaruja, Brasil, 2010, Vol. 1, pp. 115-163.
- Hsuan Y G. Temperature effect on the stress crack resistance of high-density polyethylene geomembranes. 6th International Conference on Geosynthetics, Atlanta, GA, USA, 1998, pp 371-374.
- Hsuan Y G, Koerner R M. Antioxidant depletion lifetime in high-density polyethylene geomembranes. Journal of Geotechnical and Geoenvironmental Engineering, Vol. 124, 1998, p 532.1998.
- Hsuan Y G, Schroeder H F, Rowe K, Müller W, Greenwood J H, Cazzuffi D, Koerner R M. Long-term performance and lifetime prediction of geosynthetics, 2008. EuroGeo4, Edinburgh, UK, pp 327-366.
- Müller, W.W., Jakob, I. 2003. Oxidative resistance of high density polyethylene geomembranes, Polymer Degradation and Stability Vol. 79, 2003, pp 161-172.
- Müller W, Seeger S, Thies M , Gerloff C. Long-term shear strength of multilayer geosynthetics. 3rd European Conference on Geosynthetics, Munich, Germany, 2004, ed. Floss, R., Bräu, G., Nussbaumer, M., Laackmann, K. Pp 429-434.
- Müller W W. HDPE Geomembranes in geotechnics. Springer, Berlin, Germany, 2007.
- Needham et al, Prediction of the long term generation of defects in HDPE liners, EuroGeo 3 2004a.
- Needham, A., Gallagher, E., Peggs, I., Howe, G. & Norris, J. The likely medium to long-term generation of defects in geomembrane liners. R&D Technical Report P1-500/1/TR, UK Environmental Agency, 2004b.
- Nosko, V., Andrezal, T., Gregor, T. and Ganier, P. (1996). SENSOR Damage Detection System (DDS) - The unique geomembrane testing method; in Geosynthetics: applications, design and construction. Ed., De Groot, M. B., den Hoedt, G., Termaat, R. J., Proc., 1st European Geosynthetics Conf., EuroGeo1, Maastricht, Netherlands. Balkema, Rotterdam, pp 743-748.
- Peggs, I.D., Kanninen, M.F. HDPE Geosynthetics: Premature Failures and their Prediction. Geosynthetics International, 1995, Vol. 2. No. 1. pp 327-339.
- Peggs I. Observations and thoughts on the performance of polypropylene geomembrane liners and floating covers: towards improved specifications. 2008. EuroGeo4, Edinburgh, UK, Paper 266.

- Peggs I., Elie G. Factors that affect the stress cracking resistance of HDPE geomembrane liners: Part 1 What we know and might expect. 2008. EuroGeo4, Edinburgh, UK, Paper 267.
- Peggs I. Forensic investigation of geomembrane liner cracking: Premature failure or End of Life? 9th International Conference on Geosynthetics, Guaruja, Brazil, 2010, pp 757-761.
- Rowe, RK. Long-term performance of contaminant barrier systems. 45th Rankine Lecture. Geotechnique 55 (9), 2005, pp631-678.
- Rowe R K, Islam M Z, Hsuan Y G Leachate chemical composition effects on OIT depletion in an HDPE geomembrane Geosynthetics international Vol 15 2008 pp136-151.
- Rowe R K, Rimal S. Ageing and long-term performance of geomembrane liners. 1st Pan-American Geosynthetics Conference, Cancun, Mexico, 2008, pp 425-434.
- Sangam H P, Rowe R K. Effects of exposure conditions on the depletion of antioxidants from high-density polyethylene (HDPE) geomembranes. Canadian Geotechnical Journal, Vol. 39, 2002, pp1221-1230.
- Seeger S, Müller W. Theoretical approach to designing protection: Selecting a geomembrane strain criterion. In: Jones, DRV, Dixon, N, Smith DM, Greenwood, JR. (Eds) Geosynthetics: Protecting the Environment. Thomas Telford, London, 2003, pp137-151. Seeger S, W. Müller W, Mohr K. Long term tensile testing of polyolefin fibers in geotextiles. 7th International Conference on Geosynthetics, Nice, Delmas, Gourc & Girard (eds), 2002, Swets & Zeitlinger, Lisse, Netherlands, pp1459-1462.
- Shimaoka, T., Nakayama, H., Hirai, T., Hironaka, J., Katsumi, T., Ueda, S. Kanou, H. Evaluation of physical properties of aged geomembranes taken from landfill sites in Japan. 9th International Conference on Geosynthetics, Rio, Brazil, 2010, pp 809-812.
- Soong, T-Y, Koerner, R.M. Behavior of Waves in High Density Polyethylene Geomembranes: A Laboratory Study. Proceedings of the 11th GRI Conference topic Field Installation of Geosynthetics, Geosynthetic Institute, Folsom, PA, USA, 1997, pp 128-151.
- Yu J-Y, Feng Z-G, Zhang H-L. Ageing of polymer modified bitumen (PMB) in Polymer modified bitumen –Properties and characterisation", Edited by Tony McNally, Woodhead Publishing, Oxford, UK, 2011, 264-272 (also Airey G D, same book, chapter 8, pages 251-256).

# ATTACHMENT A
# DESIGN PROCEDURE AND DESIGN OF THE GEOSYNTHETIC

## A.1 Design Procedure

The starting point of any design activity is a need or an idea, and the end point is a product that fills the need and embodies the idea.

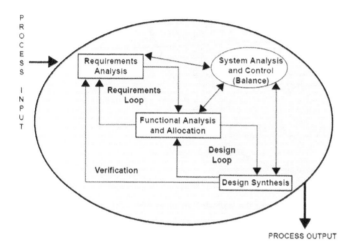

Fig. A.1 Main phases of the design Procedure.

The main phases of the design procedure (or technical process), referring to figure A.1 are:

1. Requirement analysis (clarification of the task by collecting the information about the requirements and the project constraints; requirements can be customer, functional (what), performance (how many, how far, when and how long and how often), design (e.g. how to execute), requirements )

2. Functional analysis and allocation (translation of the requirements into a coherent description of system functions that can be used to guide the Design Synthesis activity that follows.

3. Design synthesis (the search for suitable solution principles and their combinations into concept variants/design alternatives. During concept development, synthesis produces system concepts and establishes basic relationships among the subsystems. Selection of the overall topology, type of structure, and materials are some of the decisions to be made by the designer at the conceptual design phase. During preliminary and detailed design, subsystem and component descriptions are elaborated, and detailed interfaces between all system components are defined.

By means of requirements loop, the requirements will be revisited as a result of the functional analysis, caused for example by a functional conflict by mutually incompatible requirements. The design loop allows the verification that the physical design synthesized can perform the required functions at required levels of performance and help optimize the synthesized design. By means of the verification, for each application of the system engineering process, the solution will be compared to the requirements and will be verified of it works as thought it would.

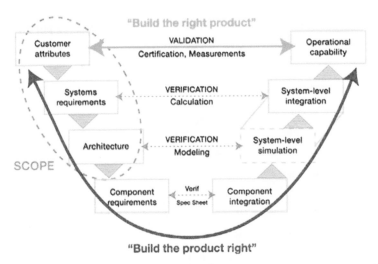

Fig. A.2 Verification and validation according to the v model for all steps in the engineering process, as it is being used by the Dutch Department of Transport (RWS) for all projects.

| |
|---|
| Verification: it works as I thought it would. |
| Validation: it looks just the real thing. |

Design factors commonly considered are:
- functional requirements
- physical constraints
- specifications
- aesthetics
- cost
- manufacturing
- evaluation/testing/analysis
- maintenance
- retirement

Detailed information about the design processes with geosynthetics can be found in several textbooks. The most important ones are listed below.

# References

*Attachment A*
* Guidelines for the design and construction of flexible revetments incorporating geotextiles for inland waterways, PIANC 1987.
* Guidelines for the design and construction of flexible revetments incorporating geotextiles in marine environment, PIANC 1992.
* V. Santvoort e.a., Geotextiles and geomembranes in civil engineering, Balkema, Rotterdam, 1994.
* R.M. Koerner, Designing with geosynthetics, Prentice-Hall, Englewood Cliffs, NJ 07632, 1987.
* P. Giroud, Geosynthetics International, Special section on Liquid Collection Systems, Vol. 7, 2000.
* Federal Waterway Engineering and Research Institute (Bundesanstalt für Wasserbau) BAW: Code of Practice Use of Geotextile Filters on Waterways (MAG), 1993.
* CUR-rapport 174, Geokunstoffen in de Waterbouw, tweede herziene uitgave, CUR, Gouda, januari 2009.
* CUR-rapport 151, Geokunststoffen in de civiele techniek. CUR, Gouda, 1991.
* CUR-rapport 217, Ontwerpen met geotextiele zandelementen, CUR, Gouda, juli 2006.
* CUR-rapport 214, Geotextiele zandelementen, CUR, Gouda, 2004.
* G.v. Santvoort, Geotextiles and Geomembranes in Civil Engineering, Balkema, Rotterdam, 1994.
* PIANC MarCom 56, The Application of Geosynthetics in Waterfront Areas, Report 113, 2011.
* CUR-rapport 206, Geokunstoffen op de bouwplaats. CUR, Gouda, 2001.
* CUR-Aanbeveling 115: 2011, Uitvoering van geokunststoffen in de waterbouw.

## References

# Attachment B
# LIST OF COMPARABLE EN, ISO, ASTM TEST METHODS

| Index Property | EN | EN | ASTM |
|---|---|---|---|
| **Physical Properties** | | | |
| Thickness | EN-1849-2 | EN ISO 9863-1 | ASTM D5199 |
| Mass per unit area | EN-1849-2 | EN ISO 9864 | ASTM D 5261 |
| | | | |
| **Hydraulic Properties** | | | |
| Characteristic opening size O90 | EN ISO 12956 | | |
| Apparent opening size | | | ASTM D4751 |
| Water permeability normal to plane V150 | EN ISO 11058 | | ASTM D4491 |
| Water permeability (liquid tightness) | pr EN 14150 | | ASTM D4491 |
| Waterflow capacity in the plane (flow rate q at specified loads and gradients) | EN ISO 12958 | | ASTM D4491 |
| Percent open area | | | COE-022125 |
| | | | |
| **Mechanical Properties** | | | |
| Tensile strength (wide strength) | EN ISO 10319 | ISO R 527 | ASTM D4595 |
| Grab Tensile Strength | EN ISO 13934-2 | | ASTM D 4632 |
| Tensile strength of seams and joints | EN ISO 10321 | | ASTM D4595 |
| Elongation (at maximum load) | EN ISO 10319 | ISO R 527 | ASTM D 4595 |
| Static puncture (CBR test) | EN ISO 12236 | | ASTM D 6241 (replaced D4833) |
| Dynamic perforation resistance (cone drop) | EN ISO 13433 (replaced EN 918) | | |
| Friction Direct shear | prEN ISO 12957-1 | | ASTM D3080 |
| Friction Inclined Plane | prEN ISO 12957-2 | | |
| Tensile Creep | EN ISO 13431 | | ASTM D6992 |

| | | | |
|---|---|---|---|
| Damage during installation | ENV ISO 10722-1 | | |
| Protection efficiency | prEN 13719:1999 | | |
| Trapezoidal tear | | | D 4533 |
| Mullen burst | | | D 3786 |
| | | | |
| **Thermal Properties** | | | |
| Low temp behaviour (flexure) | EN 495-5 | | |
| Thermal expansion | | | ASTM D 969-91 |
| | | | |
| **Durability and Chemical Resistance** | | | |
| Durability according to annex B | Annex B | | |
| Resistance to weathering | EN 12224 | | |
| UV resistance | | | ASTM D 4355 |
| Resistance to chemical aging | ENV ISO 19260 or ENV ISO 13438 | ENV 12447 | |
| Oxidation | prEN 14575 | | |
| Resistance to microbiological degradation | EN 12225 | | |
| Environmental stress cracking | | | ASTM D 5397-99 |
| Leaching (water soluble) | EN 14415 | | |
| Wetting/drying | prEN 14417 | | |
| Freezing/thawing | prEN 14418 | | |
| Root penetration | prCEN/TS 14416 | | |

# ATTACHMENT C
# STEPS TO BE TAKEN TO ASSESS THE DURABILITY OF GEOSYNTHETICS IN ACCORDANCE WITH THE LATEST REVISION OF THE EN 13249 – EN 13257 (MID 2012)

At the moment of writing, mid 2012, the CEN TC 189, is considering a change of Annex B to the EN application standards EN 13249 – EN 13257, EN 13265 and EN 15381.

It is possible that the flow scheme of steps to be taken to assess the durability of geosynthetics, which is part of annex B to the application norms will be extended.

It is therefore advised to check the latest edition of the EN application standards (EN 13249 – EN 13257, EN 13265 and EN 15381) in case fig 1.6a and b are used.

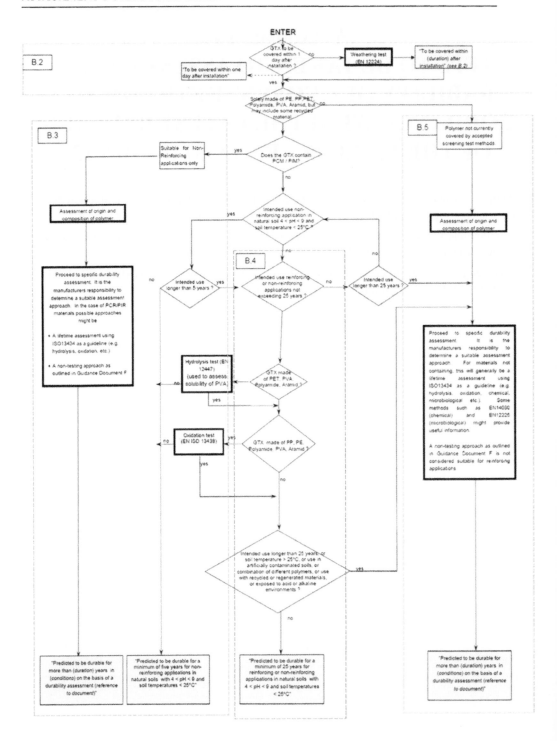

# ATTACHMENT D
# ACCREDITED INSTITUTES

The tests for the assessment of life time for a period of more than 25 years may only be executed by laboratories and institutes that are especially accredited and certified for these tests and evaluations. The accrediting and certification must be made according to EN ISO 17025 by an institute that is especially notified for that.

Certified test institutes with experience in this area are, based on the authors knowledge, For the complete list is referred to the links below:

**SKZ** -TeConA GmbH
Friedrich-Bergius-Ring 22
97076 Würzburg
Germany
www.SKZ.de

**KIWA- TBU** - Institut für Textile Bau- und Umwelttechnik GmbH
Gutenbergstr. 29
48268 Greven
Germany
www.tbu-gmbh.de or www.kiwa.de

**ITT** - Institut für Technische Textilien GmbH
im Sächsischen Textilforschungsinstitut e. V.
Annaberger Straße 240
09125 Chemnitz
Germany
www.stfi.de

**BAM** Bundesanstalt für Materialforschung und –prüfung
Fachgruppe IV.3, Abfallbehandlung und Altlastensanierung
Dr. rer. nat. Werner Müller
Unter den Eichen 87
12205 Berlin
Germany
www.bam.de

**TNO** industrie en techniek

De Rondom 1

5612 AP Eindhoven

The Netherlands

www.tno.nl

The evaluation, control, auditing and certification for durability for a period more than 25 years may only be made by an, under the CE-Mandate certified, institute. This CE-Mandate is outlined in the Construction Products Directive CPD, 89/106/EEC, 1988, amended in 1993 door Directory 93/68/EEC.

These institutes, also called "notified bodies" are identified by a number of 4 digits before the letters CPD. This number must be indicated on all certificates issued by them. The list of notified bodies is published under nr 45 (2002), C282/01 and can be found with the link: http://eur-lex.europa.eu/Result.do?aaaa=2002&mm=&jj=&type=c&nnn=282&pppp= &RechType=RECH_reference_pub&Submit=Zoeken

Google:Nando cpd› legislation› construction products› harmonized standards› EN13249:2005

Notified bodies with experience in long life time assessment of geosynthetics are:

**SKZ** - TeConA GmbH

Friedrich-Bergius-Ring 22

97076 Würzburg

Germany

**KIWA- TBU** - Institut für Textile Bau- und Umwelttechnik GmbH

Gutenbergstr. 29

48268 Greven

Germany

**ITT** - Institut für Technische Textilien GmbH

im Sächsischen Textilforschungsinstitut e. V.

Annaberger Straße 240

09125 Chemnitz

Germany

**BAM** Bundesanstalt für Materialforschung und –prüfung

Fachgruppe IV.3, Abfallbehandlung und Altlastensanierung

Dr. rer. nat. Werner Müller

Unter den Eichen 87

12205 Berlin Germany

In the UK are experienced bodies:

**BBA**,British Board of Agrement
Bucknalls Lane,

Garston, Watford, Hertfordshire,

United Kingdom, WD259BA

**BTTG**,BTTG Testing & Certification Ltd
Technology Services, Unit 4B, Stag Industrial Estate,

Atlantic Street,

Broadheath, Altrincham, Cheshire, WA14 5DW

The UK list of accredited test laboratories is on www.ukas.org. The accredited "notified" bodies can be found with the link:
http://eur-lex.europa.eu/Result.do?aaaa=2002&mm=&jj=&type=c&nnn=282&pppp=
&RechType=RECH_reference_pub&Submit=Zoeken (see also above).

Dutch notified bodies are:

**INTRON** Certificatie BV
Venusstraat, 2 Inspectie-instantie

PO Box 267, 4105 JH

NL-4100 AG Culemborg

The Netherlands

**KIWA NV**, Certificatie en keuringen
Sir Winston Churchilllaan, 273

Postbus 70

NL-2280 Ab

Rijswijk The Netherlands

The last 2 institutes do not have experience with evaluation and certification of geosynthetics above 25 years, based on our knowledge.
It is advised to ask for detailed information and confirmation about their experience in case these institutes will be used for this purpose.

# ATTACHMENT E
# QUALITY CONTROL ACCORDING TO HARMONISED EUROPEAN EN STANDARDS

## E.1 General

The quality control of geosynthetics in Europe is regulated by various EN standards and consists of various steps:

1. Quality control by the producers during production. This is laid down in procedures which may be approved by a certification institute under ISO 9001 and ISO 9002.
2. Based on this Factory Production Control System, the producer specifies the value for various properties of the material in his product data sheets.
3. According to the CEN Attestation of Conformity system 2+, a certified "notified body" institute inspects controls and certifies this Factory Production Control System. This is declared by a Factory Production Control Certificate which is issued by the "notified body" to the Producer.
4. Before the "notified body" can issue a Factory Production Control Certificate, initial inspection of the quality control system is made, including an evaluation of the Factory Production Control System. The Factory Production Control System is audited once a year and at that time it is checked if the declared properties of the products are in accordance with the results of the Factory Production Control System. If necessary the values of the properties are adjusted.
5. The producer must draw up a declaration of conformity in which he declares that the product is produced conform to the specified European Norm (e.g. EN 13249:2005)
6. Based on the Factory Production Control Certificate and the Declaration of Conformity, the producers must mark the products with a CE-Marking and deliver the products with on a Accompanying Document which gives the values for the harmonized characteristics, together with the tolerances for these values.

These various steps are shown in figure E.1.

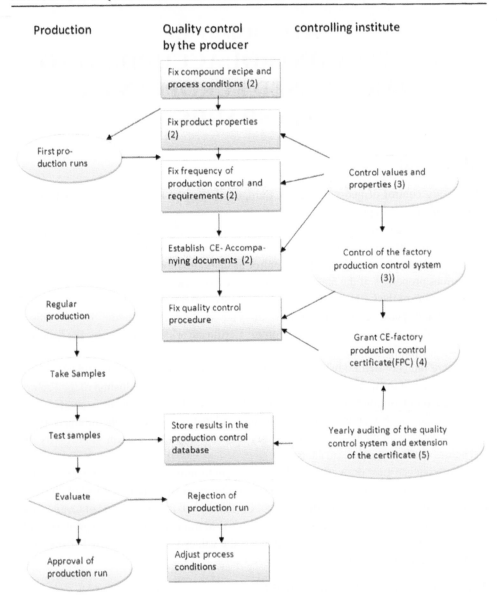

Fig. E.1 Steps in the production quality control system under CEN regulations, the numbers between brackets refer to the points in section E.1

# E.2 Initial type testing

Initial type testing is executed by the manufacturer to determine the values of the properties of a product. It is done on a number of samples, preferably from different production runs, to allow statistical evaluation. Based on the results of these tests, the values and if necessary the tolerances on the values are fixed. Instead of quoted tolerances the results can also be fixed by mean values and standard deviation. It is normally done in case new products are brought on the market or when the composition of the polymer changes.

Sometimes initial type testing can also be part of a certification process and is then executed under responsibility of the certification body.

The durability aspects of the product are also assessed as part of the initial type testing. The European application standards (EN 13249 – EN 13257, EN 13265 and EN 15381) specify that for service life times of minimum 25 years the assessment is based on screening tests according to ENV 12447 in case of polyester and ENV ISO 13438 in case of polypropylene and polyethylene. For declared service lives longer than 25 years or in case of some specific conditions, a specific durability assessment must be made. This is described in section 1.7.4.

# E.3 Factory production control

To ensure that the products have a constant quality and that the properties are the same as given in the product data sheets, a production quality control system is required. This system consists of a permanent internal production control system to ensure that all products meet the requirements. It describes the production conditions, the required checks during production, the registration of the production data directly related to roll and batch numbers. It also describes in procedures the number of samples per batch of the produced product that must be tested and the properties which must be tested. All process conditions are registered. All test results are stored in a central database, with a direct link to roll and batch numbers.

The system also specifies what action must be taken in case a product does not fulfil the requirements.

This quality control system is part of the manufacturing system conforming EN ISO 9001 or 9002.

In case a product is produced to conform to the European application standards (EN 13249 – EN 13257, EN 13265 and EN 15381), the factory production control is audited regularly and the "notified body" approves the system by the issue of a Factory Production Control Certificate. Based on this FPC certificate and with a declaration of the values of relevant properties, the European products are marked with a CE Marking.

Examples of a Factory production Certificate, a CE-Marking, a CE-Accompanying Document, a Declaration of Conformity and a CE data sheet are given in Attachment F.

## E.4 CE-Marking and CE-declaration

When the Factory Production Quality Control has been successfully made and the values of the properties which are declared by the producer have been accepted based on the initial type testing results by the certifying institute, the products must be marked with a CE-Marking. An example of such a marking on a label is given in Attachment F. The producer must also prepare a CE - Letter of conformity, in which he states what the intended use of the product is and which function is can fulfil. An example of a letter of conformity is given in Attachment F. The values of the properties which are required by the application norms, are given in a CE – Accompanying document and often also in a CE – product data sheet. Examples of these documents are also given in Attachment F.

On these various documents a statement must be made with regards to the durability of the product.
In the EN application standards (EN 13249 – 13257, EN 13265 and EN 15381). Predicted to be durable for a minimum of 5 years for non reinforcing applications in natural soils with 4<pH<9 and soil temperatures < 25°C

- Predicted to be durable for a minimum of 25 years for applications in natural soils with 4<pH<9 and soil temperatures < 25°
- In case the polymer contains recycled Post Consumer Material (PYM)/ Post Industrial Material (PIM), a special assessment must be made
- In case the intended service life is > 25 years or in soils pH<4 or pH> 9 or with soil temperature > 25°C, a special assessment must be made.

Also the time in which the product must be covered after installation is indicated on the CE documents. The categories are, based on the results of the weathering tests:
- To be covered within 1 day after installation
- To be covered within 2 weeks after installation
- To be covered within 1 month after installation.

## E.5 Quality control on site

The system for the attestation of conformity according to the application standards (EN 13249 – 13257, EN 13265 and EN 15381) indicates that the geosynthetics fall under classification 2+, which is indirect in terms of the validation of the properties of the supplied products. The product properties are verified during the initial type testing, but the values of a supplied product are only checked indirectly by an auditing institute, which audits the quality control system and database of the producer once a year.

Therefore it may be necessary to check the supplied materials after delivery on site carefully on:
- Identification of the delivered product
- Check the suitability of the product
- Check the storage- and transport facilities

- Take two samples (A and B)and test the supplied materials before installation
  - Taking samples and testing the properties after supply and before installation can be omitted, if the product is certified on the basis of a voluntary certification system. The conformity assessment institute must have been accredited and the certification system must include: initial type testing; initial inspection of the production facility; continuous surveillance, assessment and approval of the production control; audit testing of samples taken in the factory or at construction sites (CEN/TR 15019 section 4.5.1).
  - Chapter 5.2 of the application standards (EN 13249 – 13257, EN 13265 and EN 15381) describes the procedure for the testing of the two samples. If the results of sample A for a given characteristic fall within the tolerance value as declared the product is accepted, if the results fall outside 1.5 times the tolerance value the product is rejected, but the results fall in between 1 and 1.5 times the tolerance value the second sample is to be tested for that characteristic. If the results of sample B are within the limits of the tolerance value the product is accepted, otherwise it is rejected.

CEN/TR 15019 describes the actions which must be taken to control the quality of the supplied materials.

A flow sheet of these steps is given in figure E.2.

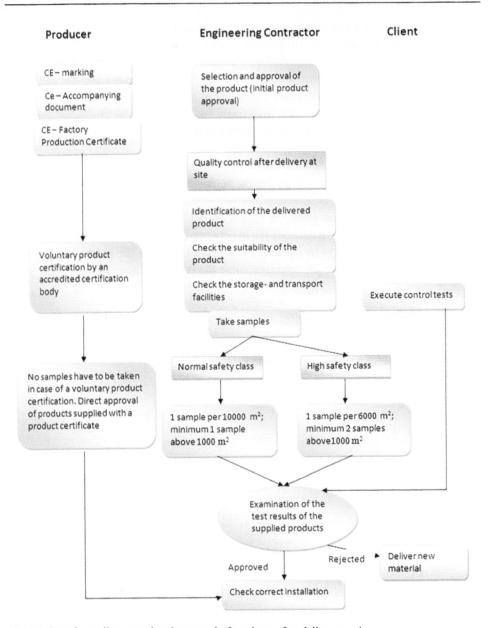

Fig. E.2 Steps in quality control and approval of products after delivery at site.

# E.6 EN – assessment of service life of geosynthetic elements

It will be clear that one of the major requirements to geosynthetics is that the geosynthetic element will be able to perform its function during the entire service life of the structure and under all conditions of loading. The service life is defined in the functional requirements of the main structure and as a consequence of that also defined as a requirement for the geosynthetic element. The conditions of loading are determined during the design of the main structure, but could also be a result from the selected installation method. **Therefore the assessment of the service life of a geosynthetics is very important.**

A method of assessment is described in the EN application standards (EN 13249 – EN 13257, EN 13265 and EN 15381). However the selection of the evaluation methods for a service life of more than 25 years is not yet normalised and left to assessment experts. Basically the method to determine the service life of a geosynthetic up to 25 years is followed, but the oxidation and hydrolysis effects are measured in a much more sophisticated way. Also the effects of various mechanisms on each other are taken into account. It will be clear that only expert institutes can perform the necessary testing and make the assessment. Therefore it is recommended that only accredited laboratories and experts are executing this work. In Europe the institutes that may perform the tests must be accredited under EN ISO 17025.

The evaluation, control and certification of products for service lives up to 25 years may according to the European Construction Products Directive CPD 89/106/EEC, 1988; amended by Directory 93/68/EEC 1993 be executed by "notified body" institutes. A list of these institutes can be found on

http://eur.lex.europa.eu/Result.do?aaaa=2002&mm=&jj=&type=c&nnn=282&pppp=&RechType=RECH_reference_pub&Submit=Zoeken.

Or google: Nando cpd› legislation› construction products› harmonized standards› EN13249:2005

More information on institutes in the Netherlands, Germany and the UK can be found in Attachment D

# E.7 Effects of the change from the EU Construction Products Directive to EU Construction Products Regulation, on the requirements in the European application standards (EN 13249 – EN 13257, EN 13265 and EN 15381), effective 1 July 2013, with regard to durability

As described above, a method of durability assessment is described in the EN application standards (EN 13249 – EN 13257, EN 13265 and EN 15381) up till a service life of minimum 25 years. Also guidelines are given on the durability assessment for longer service lives. These standards are referred to in the various chapters.

The change of the regulation in Europe will affect the EN application standards (EN 13249 – EN 13257, EN 13265 and EN 15381). They are in the process of being revised. The major differences are expected to be:

- The tables 1 in each application standard which describes the functions, function-related characteristics and test method used, for the relevant application will be changed. In general the tests indicated as relevant to all conditions of use en those indicated as relevant to specific condition of use, could become required tests, which means that the values for these characteristics must be declared by the manufacturer in de accompanying documents.
- The manufacturer must declare the performance (value) of the product for the characteristic, instead of declare the conformity of the product. Declaration means that the manufacturer declares the value of the characteristic, (possibly, but not necessarily) based on his internal quality control system. A factory production control is executed by a "notified" body to approve and audit the quality control system of the manufacturer. A Certificate of Consistency of Performance will be issued by the "notified" body after the regulation becomes effective in 1 July 2013.
- The new standards will also set a minimum frequency for each test, including the durability screening tests.
- It is the intention of the EU to include the durability of the characteristic into these standards (i.e. the durability of permeability, opening size, etc) and not limit the durability to the polymer alone. However, these tests have not been developed yet.

It is not clear at the time of writing, what the exact changes to the application standards (EN 13249 – EN 13257, EN 13265 and EN 15381) will be. It is therefore advised to check the actual version of the standard when applicable.

A reference is made to the EU Construction Products Regulation, Regulation (EU) No 305/2011 of the European Parliament and of the Council laying down harmonised conditions for the marketing of construction products and repealing Council Directive 89/106/EEC.;http://www.europarl.europa.eu/oeil/file.jsp?id=5642292.

# ATTACHMENT F
# EXAMPLES OF CE DOCUMENTS

CE – Accompanying document

0799-CPD

## Colbonddrain CX1000

Colbond bv, Westervoortsedijk 73,
6827 AV Arnhem, The Netherlands

02

0799-CPD-11

Geocomposite for application

EN 13249   EN 13250   EN13251   EN 13252   EN 13253   EN 13254   EN 13255   EN 13257   EN 13265

Intended uses: F, F + S, F + S + D

Filtration    Separation    Drainage

| Properties | Test method | Mean value | Tolerance |
|---|---|---|---|
| Tensile strength | EN ISO 10319 | MD  20.0 kN/m<br>CMD  20.0 kN/m | - 2.0 kN/m<br>- 2.0 kN/m |
| Resistance to static puncture | EN ISO 12236 | 1.3 kN | - 0.2 kN |
| Dynamic perforation resistance | EN 918 | 30 mm | + 5 mm |
| Characteristic opening size | EN ISO 12956 | 72 μm | +/- 7 μm |
| Water permeability, $VI_{H50}$ | EN ISO 11058 | 70 mm/s | - 10 mm/s |
| Water flow capacity in the plane | EN ISO 12958<br>350 kPa, i=1.0, f/f | $0.85 \cdot 10^{-3}$ m²/s | $- 0.15 \cdot 10^{-3}$ m²/s |
| Durability | - to be covered within 2 weeks after installation<br>- predicted to be durable for a minimum of 5 years in<br>   natural soils with 4<pH<9 and soil temperature <25°C | | |

AD-ED-278-GB-A-10/2002

CE – Accompanying document with the values for the properties which must be declared according to the application norms EN 13249 – 13255. Further the intended use is indicated and the application of the geosynthetic.

CE- product information sheet

# COLBOND
### GEOSYNTHETICS

## Colbonddrain® CX1000
### CE Product Information

### Product description

Colbonddrain CX1000 consists of a black three-dimensional polyester filament core structure wrapped in a natural colored, nonwoven polyester needle punched adhesive bonded filter. The filter and core are ultrasonically welded together at the edges to produce a fully integrated product that prevents easy intrusion of the filter into the core structure.

### Application

Colbonddrain CX1000 is usually used for the vertical drainage of soft clay/silt soils typically in reclamation projects.

### Performances

**Properties of the composite**

| Applied stress in kPa | Hydraulic gradient i in - | Water flow capacity in-plane* $q_{stress/gradient}$ in  l/(s·m)** | | |
|---|---|---|---|---|
| | | Mean value | Tolerance value | |
| 350 | 1.0 | 0.85 | -0.15 | EN ISO 12958 |

* Test results by Colbond Geosynthetics Laboratory according to EN ISO 12958, opt. F/F.
  Soil pressure against both filter sides was simulated by foam layers.
** l/(s·m) = $10^{-3}$ m²/s

**Hydraulic properties of the filter**

| | | Mean value | Tolerance value | |
|---|---|---|---|---|
| Water permeability, VI$_{H50}$ | mm/s | 70 | - 10 | EN ISO 11058 |
| Opening size | µm | 72 | +/- 7 | EN ISO 12956 |

**Mechanical properties of the filter**

| | | Mean value | Tolerance value | |
|---|---|---|---|---|
| Polymer | | PET | | |
| Mass per unit area | g/m² | 160 | - 20 | EN 965 |
| Thickness | mm | 0.6 | - 0.1 | EN 964-1 |
| Tensile strength MD | kN/m | 10 | - 1 | EN 10319 |
| Tensile strength CMD | kN/m | 10 | - 1 | EN 10319 |
| Elongation at break MD | % | 37 | - 6 | EN 10319 |
| Resistance to static puncture | kN | 1.3 | - 0.2 | EN 12236 |
| Dynamic perforation resistance | mm | 30 | + 5 | EN 918 |

Mean value + tolerance value correspond to the 95% confidence level.

Page 1 of a CE – product information sheet, this is a product information sheet with values for the properties which are required in the relevant CE application norms, extended with some other relevant properties

## Dimensions and weights

| Colbonddrain Vertical Drain | Mattings | | | | | Rolls | | |
|---|---|---|---|---|---|---|---|---|
| Type | Thickness | Weight | Width | Length | Area | Ø | Length | Gross-weight |
| | mm | g/m² | m | m | m² | m | m | kg |
| CX1000 | 5 | 700 | 0.1 | 200 | 20 | 1.20 | 0.30 | 42 |

Individual values may vary from above mentioned data

## Quality Assurance

The Quality Management System of Colbond Geosynthetics, at Arnhem (development and sales) and Obernburg (production), has been approved by Lloyd's Register Quality Assurance Limited for the ISO 9001:2000 quality management system standard (Certificate No 935136)

CE
0799-CPD

Colbonddrain CX1000 is CE-certified by an independent notified body (0799-CPD)

Colbond Geosynthetics, P.O. Box 9600, 6800 TC Arnhem, the Netherlands
Phone: +31 26 366 4600 • Fax: +31 26 366 5812
geosynthetics@colbond.com • www.colbond.com

Page 2 of a CE – product information sheet, this is a product information sheet with values for the properties which are required in the relevant CE application norms, extended with some other relevant properties

CE – Factory production control certificate

Institut für textile Bau- und Umwelttechnik GmbH
Institute for textile building and environment technology
Gutenbergstr. 29 • 48268 Greven • Germany

## Certificate 0799-CPD-10

### Factory Production Control

In compliance with the Directive 89/106/EEC of the Council of European Communities of 21 December 1988 on the approximation of laws, regulations and administrative provisions of the Member States relating to the construction products (Construction Products Directive – CPD), amended by the Directive 93/68/EEC of the Council of European Communities of 22 July 1993, it has been stated that the construction products

propex®
60-6054, 60-6055, 60-6056, 60-6057, 60-6060, 60-6061, 60-6081, 60-6082, 60-6083, 60-6084, 60-6156, 60-7030, 60-7041, 60-7050, 60-7060, 60-7063, 60-7080, 60-7087, 60-7100, 60-7120, 60-7121, 60-7150, 60-7061, 60-6088, 60-7151, 60-7040, 60-7081;

61-6083, 61-6084, 61-7060, 61-7080, 61-7100, 61-7087, 61-6082, 61-7041;

65-6055, 65-6057, 65-6083, 65-6085, 65-6088, 65-7040, 65-7060, 65-7087, 65-7100

produced by the manufacturer

### PROPEX Fabrics GmbH
Düppelstraße 16
48599 Gronau
Germany

in the factory        Code 10440-A

are submitted by the manufacturer to the initial type-testing of the products, a factory production control and to the further testing of samples taken at the factory in accordance with a prescribed test plan and that the notified body tBU - Institut für textile Bau- und Umwelttechnik GmbH, Greven, Germany, has performed the initial inspection of the factory and of the factory production control and performs the continuous surveillance, assessment and approval of the factory production control.

This certificate attests that all provisions concerning the attestation of factory production control described in Annex ZA of the standards

EN    13249, 13250, 13251, 13252, 13253,
13254, 13255, 13256, 13257, 13265

were applied.

This certificate was first issued on 15.09.2002 and remains valid as long as the conditions laid down in the harmonised technical specification in reference or the manufacturing conditions in the factory or the FPC itself are not modified significantly.

Greven, 10.07.2008

Prof. Dr.-Ing. Müller-Rochholz

This certificate including newest products: www.tbu-gmbh.de

CE – Factory production control certificate. This certificate states that for the mentioned product types an initial product type testing has been executed, that initial inspection of the factory quality control system has been executed and that at regular times audits are made to control the factory quality control system.

Declaration of conformity

# EC declaration of conformity

Name and address of manufacturer
Colbond bv
Westervoortsedijk 73
6800 TC Arnhem
The Netherlands

Description of the product
Geocomposite

Colbonddrain® CX 1000

Intended uses:
Filtration + Separation + Drainage (F + S + D)

Provisions to which the product conforms
For application in:

| | |
|---|---|
| Construction of roads and other trafficked areas | EN 13249 (F + S) |
| Construction of railways | EN 13250 (F + S) |
| Earthworks, foundations and retaining structure | EN 13251 (F + S) |
| Drainage systems | EN 13252 (F + S + D) |
| External erosion control systems | EN 13253 (F + S) |
| Construction of reservoirs and dams | EN 13254 (F + S) |
| Construction of canals | EN 13255 (F + S) |
| Solid waste disposals | EN 13257 (F + S) |
| Liquid waste containment projects | EN 13265 (F) |

Particular conditions applicable to the use of the product
None

Name and address of the approved body
0799-CPD
Institut für textile Bau- und Umwelttechnik GmbH
Gutenbergstrasse 29
D-48268 Greven

Name and position of signatory
Wim Voskamp
Manager R&D and Technical Marketing

9/10/02

Signature                    Date

## Example of a product label with CE Marking

# ATTACHMENT G
# RELEVANT NORMS AND STANDARDS

NEN EN ISO 10320:1999; Geotextiel en soortgelijke producten - Identificatie op de bouw-plaats.

EN 12224:2000 Geotextiles and geotextile-related products - Determination of the resistance to weathering.

EN 12225:2000 Geotextiles and geotextile-related products - Method for determining the microbiological resistance by a soil burial test.

EN 12226:2000 Geotextiles and geotextile-related products - General tests for evaluation following durability testing.

EN 12447:2001 Geotextiles and geotextile-related products – Screening test method for determining the resistance to hydrolysis in water.

EN 13719:2002 Geotextiles and geotextile-related products - Determination of the long term protection efficiency of geotextiles in contact with geosynthetic barriers.

EN 13738:2004 Geotextiles and geotextile-related products – Determination of pullout resistance in soil.

EN 14030:2001 Geotextiles and geotextile-related products – Screening test method for determining the resistance to acid and alkaline liquids (ISO/TR 12960:1998, modified).

EN 14574:2004 Geosynthetics - Determination of the pyramid puncture resistance of supported geosynthetics.

EN 14575:2005 Geosynthetic barriers - Screening test method for determining the resistance to oxidation.

EN ISO 9862:2005 Geosynthetics - Sampling and preparation of test specimens (ISO 9862:2005).

EN ISO 9863-1:2005 Geosynthetics - Determination of thickness at specified pressures - Part 1: Single layers (ISO 9863-1:2005).

EN ISO 9863-2:1996 Geotextiles and geotextile-related products – Determination of thickness at specified pressures - Part 2: Procedure for determination of thickness of single layers of multilayer products (ISO 9863-2:1996).

EN ISO 9864:2005 Geosynthetics - Test method for the determination of mass per unit area of geotextiles and geotextile-related products (ISO 9864:2005).

EN ISO 10318:2005 Geosynthetics - Terms and definitions (ISO 10318:2005).

EN ISO 10319:2008 Geosynthetics - Wide-width tensile test (ISO 10319:2008).

EN ISO 10320:1999 Geotextiles and geotextile-related products - Identification on site (ISO 10320:1999).

EN ISO 10321:2008 Geosynthetics - Tensile test for joints/seams by wide-width strip method (ISO 10321:2008).

EN ISO 10722:2007 Geosynthetics - Index test procedure for the evaluation of mechanical damage under repeated loading - Damage caused by granular material (ISO 10722:2007).

EN ISO 11058:1999 Geotextiles and geotextile-related products - Determination of water permeability characteristics normal to the plane, without load (ISO 11058:1999). EN ISO 12236:2006 Geosynthetics - Static puncture test (CBR test) (ISO 12236:2006).

EN ISO 12956:1999 Geotextiles and geotextile-related products - Determination of the characteristic opening size (ISO 12956:1999).

EN ISO 12957-1:2005 Geosynthetics - Determination of friction characteristics - Part 1: Direct shear test (ISO 12957-1:2005).

EN ISO 12957-2:2005 Geosynthetics - Determination of friction characteristics - Part 2: Inclined plane test (ISO 12957-2:2005).

EN ISO 12958:1999 Geotextiles and geotextile-related products - Determination of water flow capacity in their plane (ISO 12958:1999).

EN ISO 13426-1:2003 Geotextiles and geotextile-related products - Strength of internal structural junctions - Part 1: Geocells (ISO 13426- 1:2003).

EN ISO 13426-2:2005 Geotextiles and geotextile-related products - Strength of internal structural junctions - Part 2: Geocomposites (ISO 13426-2:2005).

EN ISO 13427:1998 Geotextiles and geotextile-related products – Abrasion damage simulation (sliding block test) (ISO 13427:1998).

EN ISO 13428:2005 Geosynthetics - Determination of the protection efficiency of a geosynthetic against impact damage (ISO 13428:2005).

EN ISO 13431:1999 Geotextiles and geotextile-related products – Determination of tensile creep and creep rupture behaviour (ISO 13431:1999).

EN ISO 13433:2006 Geosynthetics - Dynamic perforation test (cone drop test) (ISO 13433:2006).

EN ISO 13437:1998 Geotextiles and geotextile-related products - Method for installing and extracting samples in soil, and testing specimens in laboratory (ISO 13437:1998).

EN ISO 13438:2004 Geotextiles and geotextile-related products – Screening test method for determining the resistance to oxidation (ISO 13438:2004).

EN ISO 25619-1:2008 Geosynthetics - Determination of compression behaviour - Part 1: Compressive creep properties (ISO 25619-1:2008).

EN ISO 25619-2:2008 Geosynthetics - Determination of compression behaviour - Part 2: Determination of short-term compression behavior (ISO 25619-2:2008).

ISO/TR 13434:2008 Geosynthetics -- Guidelines for the assessment of durability.

ISO/TR 15010:2005 Geosynthetics – Quality control at the project site.

EN 12224:2000 – Geotextiles and geotextile-related products – Determination of the resistance to weathering.

EN 12225:2000 – Geotextiles and geotextile-related products – Method for determining the microbiological resistance by a soil burial test.

EN 12226:2000 – Geotextiles and geotextile-related products – General tests for evaluation following durability testing.

EN 12447:2001 – Geotextiles and geotextile-related products – Screening test method for determining the resistance to hydrolysis in water.

EN 14030:2001 – Geotextiles and geotextile-related products – Screening test method for determining the resistance to acid and alkaline liquids.

EN ISO 13438:2004 – Geotextiles and geotextile-related products – Screening test method for determining the resistance to oxidation. E

N 13249:2000: 2001, Geotextiles and geotextile-related products – Characteristics required for use in the construction of roads and other trafficked areas (excluding railways and asphalt inclusion).

EN 13250:2001; Geotextiles and geotextile-related products. Characteristics required for use in the construction of railways.

EN 13251:2001; Geotextiles and geotextile-related products. Characteristics required for use in earthworks, foundations and retaining structures.

EN 13252:2001; Geotextiles and geotextile-related products. Characteristics required for use in drainage systems.

EN 13253:2001, Geotextiles and geotextile-related products. Characteristics required for use in erosion control works (coastal protection, bank revetments).

EN 13254: 2001 Geotextiles and geotextile-related products - Characteristics required for the use in the construction of reservoirs and dams.

EN 13255: 2000, Geotextiles and geotextile-related products - Characteristics required for use in the construction of canals.

EN 13256: 2001 Geotextiles and geotextile-related products - Characteristics required for use in the construction of tunnels and underground structures.

EN 13257: 2001 Geotextiles and geotextile-related products. Characteristics required for use in solid waste disposals.

Printed and bound by CPI Group (UK) Ltd, Croydon, CR0 4YY

24/10/2024

01778292-0003